Antarctic Penguins

A Study of Their Social Habits

G. Murray Levick

Alpha Editions

This edition published in 2024

ISBN : 9789367248201

Design and Setting By
Alpha Editions
www.alphaedis.com
Email - info@alphaedis.com

As per information held with us this book is in Public Domain.
This book is a reproduction of an important historical work. Alpha Editions uses the best technology to reproduce historical work in the same manner it was first published to preserve its original nature. Any marks or number seen are left intentionally to preserve its true form.

ADÉLIE PENGUINS(1)

INTRODUCTION

THE penguins of the Antarctic regions very rightly have been termed the true inhabitants of that country. The species is of great antiquity, fossil remains of their ancestors having been found, which showed that they flourished as far back as the eocene epoch. To a degree far in advance of any other bird, the penguin has adapted itself to the sea as a means of livelihood, so that it rivals the very fishes. This proficiency in the water has been gained at the expense of its power of flight, but this is a matter of small moment, as it happens.

In few other regions could such an animal as the penguin rear its young, for when on land its short legs offer small advantage as a means of getting about, and as it cannot fly, it would become an easy prey to any of the carnivora which abound in other parts of the globe. Here, however, there are none of the bears and foxes which inhabit the North Polar regions, and once ashore the penguin is safe.

The reason for this state of things is that there is no food of any description to be had inland. Ages back, a different state of things existed: tropical forests abounded, and at one time, the seals ran about on shore like dogs. As conditions changed, these latter had to take to the sea for food, with the result that their four legs, in course of time, gave place to wide paddles or "flippers," as the penguins' wings have done, so that at length they became true inhabitants of the sea.

Were the Sea-Leopards(2) (the Adélies' worst enemy) to take to the land again, there would be a speedy end to all the southern penguin rookeries. As these, however, are inhabited only during four and a half months of the year, the advantage to the seals in growing legs again would not be great enough to influence evolution in that direction. At the same time, I wonder very much that the sea-leopards, who can squirm along at a fair pace on land, have not crawled up the few yards of ice-foot intervening between the water and some of the rookeries, as, even if they could not catch the old birds, they would reap a rich harvest among the chicks when these are hatched. Fortunately however they never do this.

FIG. 1. AN ANGRY ADÉLIE

When seen for the first time, the Adélie penguin gives you the impression of a very smart little man in an evening dress suit, so absolutely immaculate is he, with his shimmering white front and black back and shoulders. He stands about two feet five inches in height, walking very upright on his little legs.

His carriage is confident as he approaches you over the snow, curiosity in his every movement. When within a yard or two of you, as you stand silently watching him, he halts, poking his head forward with little jerky movements, first to one side, then to the other, using his right and left eye alternately during his inspection. He seems to prefer using one eye at a time when viewing any near object, but when looking far ahead, or walking along, he looks straight ahead of him, using both eyes. He does this, too, when his anger is aroused, holding his head very high, and appearing to squint at you along his beak, as in Figure 1.

After a careful inspection, he may suddenly lose all interest in you, and ruffling up his feathers sink into a doze. Stand still for a minute till he has settled himself to sleep, then make sound enough to wake him without startling him, and he opens his eyes, stretching himself, yawns, then finally walks off, caring no more about you. (Figs. 2 and 3.)

The wings of Adélies, like those of the other penguins, have taken the form of paddles, and are covered with very fine scale-like feathers. Their legs being very short, they walk slowly, with a waddling gait, but can travel at a fair pace over snow or ice by falling forward on to their breasts, and propelling themselves with all four limbs.

To continue the sketch, I quote two other writers:

M. Racovitza, of the "Belgica" expedition, well describes them as follows:

"Imagine a little man, standing erect, provided with two broad paddles instead of arms, with head small in comparison with the plump stout body; imagine this creature with his back covered with a black coat ... tapering behind to a pointed tail that drags on the ground, and adorned in front with a glossy white breast-plate. Have this creature walk on his two feet, and give him at the same time a droll little waddle, and a pert movement of the head; you have before you something irresistibly attractive and comical."

Fig. 2. Dozing

FIG. 3. WAKING UP, STRETCHING, AND YAWNING

Dr. Louis Gain, of the French Antarctic expedition, gives us the following description:

"The Adélie penguin is a brave animal, and rarely flees from danger. If it happens to be tormented, it faces its aggressor and ruffles the black feathers which cover its back. Then it takes a stand for combat, the body straight, the animal erect, the beak in the air, the wings extended, not losing sight of its enemy.

"It then makes a sort of purring, a muffled grumbling, to show that it is not satisfied, and has not lost a bit of its firm resolution to defend itself. In this guarded position it stays on the spot; sometimes it retreats, and lying flat on the ground, pushes itself along with all the force of its claws and wings. Should it be overtaken, instead of trying to increase its speed, it stops, backs up again to face anew the peril, and returns to its position of combat. Sometimes it takes the offensive, throws itself upon its aggressor, whom it punishes with blows of its beak and wings."

The Adélie penguin is excessively curious, taking great pains to inspect any strange object he may see. When we were waiting for the ship to fetch us home, some of us lived in little tents which we pitched on the snow about fifty yards from the edge of the sea. Parties of penguins from Cape Royds rookery frequently landed here, and almost invariably the first thing they did on seeing our tents, was at once to walk up the slope and inspect these, walking all round them, and often staying to doze by them for hours. Some of them, indeed, seemed to enjoy our companionship. When you pass on the sea-ice anywhere near a party of penguins, these generally come up to look at you, and we had great trouble to keep them away from the sledge dogs when these were tethered in rows near the hut at Cape Evans. The dogs killed large numbers of them in consequence, in spite of all we could do to prevent this.

The Adélies, as will be seen in these pages, are extremely brave, and though panic occasionally overtakes them, I have seen a bird return time after time to attack a seaman who was brutally sending it flying by kicks from his sea-boot, before I arrived to interfere. An exact description of the plumage of the Adélie penguins will be found in the Appendix, as it is more especially of their habits that I intend to treat in this work.

Before describing these, and with a view to making them more intelligible to the general reader, I will proceed to a short explanation.

The Adélie penguins spend their summer and bring forth their young in the far South. Nesting on the shores of the Antarctic continent, and on the islands of the Antarctic seas, they are always close to the water, being dependent on the sea for their food, as are all Antarctic fauna; the frozen regions inland, for all practical purposes, being barren of both animal and vegetable life.

Their requirements are few: they seek no shelter from the terrible Antarctic gales, their rookeries in most cases being in open wind-swept spots. In fact, three of the four rookeries I visited were possibly in the three most windy regions of the Antarctic. The reason for this is that only wind-swept places are so kept bare of snow that solid ground and pebbles for making nests are to be found.

When the chicks are hatched and fully fledged, they are taught to swim, and when this is accomplished and they can catch food for themselves, both young and old leave the Southern limits of the sea, and make their way to the pack-ice out to the northward, thus escaping the rigors and darkness of the Antarctic winter, and keeping where they will find the open water which they need. For in the winter the seas where they nest are completely covered by a thick sheet of ice which does not break out until early in the following summer. Much of this ice is then borne northward by tide and wind, and accumulates to form the vast rafts of what is called "pack-ice," many hundreds of miles in extent, which lie upon the surface of the Antarctic seas. (Fig. 4.)

It is to this mass of floating sea-ice that the Adélie penguins make their way in the autumn, but as their further movements here are at present something of a mystery, the question will be discussed at greater length presently.

When young and old leave the rookery at the end of the breeding season, the new ice has not yet been formed, and their long journey to the pack has to be made by water, but they are wonderful swimmers and seem to cover the hundreds of miles quite easily.

Arrived on the pack, the first year's birds remain there for two winters. It is not until after their first moult, the autumn following their departure from the rookery, that they grow the distinguishing mark of the adult, black feathers replacing the white plumage which has hitherto covered the throat.

The spring following this, and probably every spring for the rest of their lives, they return South to breed, performing their journey, very often, not only by water, but on foot across many miles of frozen sea.

Fig. 4. Pack Ice (on which the Adélies winter)
Two Weddell Seals are seen on a Floe

Fig. 5. Heavy Seas in the Autumn

For those birds who nest in the southernmost rookeries, such as Cape Crozier, this journey must mean for them a journey of at least four hundred miles by water, and an unknown but considerable distance on foot over ice.

As I am about to describe the manners and customs of Adélie penguins at the Cape Adare rookery, I will give a short description of that spot.

Cape Adare is situated in lat. 71° 14' S. long. 170° 10' E., and is a neck of land jutting out from the sheer and ice-bound foot-hills of South Victoria Land northwards for a distance of some twenty miles.

For its whole length, the sides of this Cape rise sheer out of the sea, affording no foothold except at the extreme end, where a low beach has been formed, nestling against the steep side of the cliff which here rises almost perpendicularly to a height of over 1000 feet.

Hurricanes frequently sweep this beach, so that snow never settles there for long, and as it is composed of basaltic material freely strewn with rounded pebbles, it forms a convenient nesting site, and it was on this spot that I made the observations set forth in the following pages.

Viewed before the penguins' arrival in the spring, and after recent winds had swept the last snowfalls away, the rookery is seen to be composed of a series of undulations and mounds, or "knolls," while several sheets of ice, varying in size up to some hundreds of yards in length and one hundred yards in width, cover lower lying ground where lakes of thaw water form in the summer. Though doubtless the ridges and knolls of the rookery owe their origin mainly to geological phenomena, their contour has been much added to as, year by year, the penguins have chosen the higher eminences for their nests; because their guano, which thickly covers the higher ground, has protected this from weathering and the denuding effect of the hurricanes which pass over it at certain seasons and tend to carry away the small fragments of ground that have been split up by the frost.

The shores of this beach are protected by a barrier of ice-floes which are stranded there by the sea in the autumn. These floes become welded together and form the "ice-foot" frequently referred to in these pages, and photographs showing how this is done are seen on Figs. 5, 6, 7 and 8.

At the back of the rookery, nesting sites are to be seen stretching up the steep cliff to a height of over 1000 feet, some of them being almost inaccessible, so difficult is the climb which the penguins have made to reach them.

Fig. 6. "… Throw up Masses of Ice,

FIG. 7. "… WHICH ARE FROZEN INTO A COMPACT MASS AS WINTER APPROACHES"

On Duke of York Island, some twenty miles south of the Cape Adare rookery, another breeding-place has been made. This is a small colony only, as might be expected. Indeed it is difficult to see why the penguins chose this place at all whilst room still exists at the bigger rookery, because Duke of York Island, until late in the season, is cut off from open water by many miles of sea-ice, so that with the exception of an occasional tide crack, or seals' blow holes, the birds of that rookery have no means of getting food except by making a long journey on foot. When the arrivals were streaming up to Cape Adare many were seen to pass by, making in a straight line for Duke of York Island, and so adding another twenty miles on foot to the journey they had already accomplished.

When the time arrived for the birds to feed, some open leads had formed about half way across the bay, and those of the Duke of York colony were to be seen streaming over the ice for many miles on their way between the water and their nests. They seem to think nothing of long journeys, however, as in the early season, when unbroken sea-ice intervened between the two rookeries, parties of penguins from Cape Adare actually used to march out and meet their Duke of York friends half way over, presumably for the pleasure of a chat.

To realize what this meant, we must remember that an Adélie penguin's eyes being only about twelve inches above the ground when on the march, his horizon is only one mile distant. Thus from Cape Adare he could just see the top of the mountain on Duke of York Island peeping above the horizon *on the clearest day*. In anything like thick weather he could not see it at all, and probably he had never been there. So in the first place, what was it that impelled him to go on this long journey to meet his friends, and when so impelled, what instinct pointed out the way? This of course merely brings us to the old question of migratory instinct, but in the case of the penguin, its horizon is so very short that it is quite evident he possesses a special sense of direction, in addition to the special sense which urged him to go and meet the Duke of York Island contingent, and I may here remark that when we were returning to New Zealand in the summer of 1913, we passed troops of penguins swimming in the open sea far out of sight of land,—an unanswerable reply to those naturalists who still maintain that migrating birds must rely upon their eyes for guidance, and this remark applies equally to the penguins we found on the northern limits of the pack-ice, some five hundred miles from the rookeries to which they would repair the following year.

TABLE A						
Mean date	Northern limit of pack	Miles from C. Adare	Southern limit of pack	Miles of pack N. and S.	Remarks
Feb. 3, 1839	68° S.	190	?	?	Balleny
Jan. 1, 1841	66° 30'	280	69°	150	Ross
Feb. 1, 1895	66° 15'	300	69° 45'	210	Kristensen
Feb. 8, 1899	66° 0'	315	69° 0'	180	Borchgravink
Feb. 27, 1904	?		70° 30'	?	Scott
Feb. 15, 1910	nil		nil		*Terra Nova*
Mar. 13, 1912	nil		nil		*Terra Nova*
Jan. 30, 1913	nil		nil		*Terra Nova*

Note.—Ross, Kristensen, Scott, Shackleton and Pennell all, however, found pack late in the season while trying to work west along the coast when only some forty-five to seventy-five miles north of Cape Adare, and all were turned by this pack.

According to Pennell, it appears probable that there is a great hang of pack in the sea west of Cape Adare and south of the Balleny Islands, and most likely it is here that the Adélies repair when they leave Cape Adare rookery in the autumn. I think, however, it is safe to assume that they seek the northernmost limits of the pack during the winter, as these would offer the most favourable conditions.

TABLE B				
Date	Longitude	Northern limit	Extends N. and S. Miles	Minutes of latitude Northern limit is N. of Cape Adare
Jan. 12, 1840	166° E.	64° 30'	—	400 (Wilkes)
Jan. 3, 1902	178° E.	67° S.	140	250 (*Discovery*)
Dec. 31, 1902	180° E.	66° 30'	60	280 (*Morning*)
	Second belt	69°	30	130 (*Morning*)
Dec. 20, 1908	178° W.	66° 30'	60	270 (*Nimrod*)
Dec. 9, 1910	178° W.	64° 45'	300	390 (*Terra Nova*)
Dec. 27, 1911	177° W.	65° 20'	160	360 (*Terra Nova*)
Mar. 8, 1911	162° E.	64° 30'	270	400 (*Terra Nova*)

Fig. 8. "… And later, form the Beautiful Terraces of the Ice-foot"

Fig. 9. Penguins at the Rookery

The exact whereabouts of the Adélie penguins during the winter months has been much discussed by different writers. It is agreed that they repair to the pack-ice, but our knowledge of the movements of this pack is very vague at the present time, and so unfortunately I can give but a rough idea of the subject.

I have collected and noted down the latest evidence for the benefit of the zoologists of future expeditions who may wish to investigate the matter further, and I am indebted for nearly the whole of it to Commander Harry L. Pennell, R.N., commander of the *Terra Nova* from 1910 to 1913, who kindly drew up for me Tables A and B (*see* pp. 13 and 14).

Probably the information which more nearly concerns the penguins of Cape Adare rookery will be found in Table A. The birds from Cape Crozier and Cape Royds rookeries must have some four hundred miles further to travel when they go North in the autumn than those at Cape Adare.

PART I

THE FASTING PERIOD

Diary from October 13 to November 3, describing the arrival of the Adélie penguins at the rookery, and habits during the periods of mating and building.

THE first Adélie penguins arrived at the Ridley Beach rookery, Cape Adare, on October 13. A blizzard came on then, with thick drift which prevented any observations being made. The next day, when this subsided, there were no penguins to be seen.

On October 15 two of them were loitering about the beach. During the forenoon they were separate, but in the afternoon they kept company, and walked over to the south-east corner of the rookery under the cliff of Cape Adare, where they were sheltered from the cold breeze.

On October 16 at 11 A.M. there were about twenty penguins arrived. Several came singly, and one little party of three came up together. On arrival they wandered about by themselves, and stood or walked about the beach, giving one the impression of simply hanging about, waiting for something to "turn up."

By 4 P.M. there must have been close on a hundred penguins at the rookery. It was a calm day and misty, so that I could not see far out across the sea-ice, but so far it was evident that the birds were not arriving in batches, but just dribbling in. They were then for the most part squatting about the rookery, well scattered, some solitary, others in groups, and facing in all directions. (Fig. 9.) They were not on the prominences where the nesting sites are, but in the hollows and on the snow of the frozen lakes. There was no sign of love-making or any activity whatever. All were in fine plumage and condition.

During the night of October 16 the number of penguins increased greatly, and on the morning of the 17th there was a thin sprinkling scattered over the rookery. (Fig. 11.) A few were in pairs or threes, but more in groups of a dozen or more, and all the birds were very phlegmatic, many of them lying on their breasts, with beaks outstretched, apparently asleep, and nearly all, as yesterday, in the hollows, though there was no wind, and away from the nesting sites. They were very quiet. Probably they were fatigued after their journey; perhaps also they were waiting the stimulation of a greater crowd before starting their breeding operations. As the guano-covered ridges, on which the old nests are, were fairly soft and the pebbles loose, they were not waiting for higher temperatures in order to get to work.

During October 17 the arrivals became gradually more frequent. They were dribbling up from the sea-ice at the north-end of the beach, and soon made a well-worn track up the ice-foot, whilst a long line of birds approaching in single file, with some gaps, extended to the horizon in a northerly direction.

During the day I noticed some penguins taking possession of old nests on the ridges. These mostly squatted in the nests without any attempt at repairing them or rearrangement of any sort. Afterwards I found that they were unmated hens waiting for mates to come to them, and that this was a very common custom among them. (Fig. 10.) If two occupied nests within reach of one another they would stretch out their necks and peck at each other. Their endeavour seemed to be to peck each other's tongue, and this they frequently did, but generally struck the soft parts round the margin of the bills, which often became a good deal swollen in consequence. Often also their beaks would become interlocked. They would keep up this peck-pecking hour after hour in a most relentless fashion. (Fig. 12.) On one occasion I saw a hen succeed in driving another off one of the old nests which she occupied. The vanquished one squatted on the ground a few yards away, with rumpled feathers and "huffy" appearance, whilst the other walked on to the nest and assumed the "ecstatic" attitude (page 46). Nothing but animosity could have induced this act, as thousands of old unoccupied nests lay all around.

About 9 P.M. a light snowstorm came on, and those few birds who had taken possession of nests, left them, and all now lay in the hollows, nestling into the fine drift which soon covered the ground to the extent of a few inches. A group of about a dozen penguins which arrived near the ice-foot in the morning, halted on the sea-ice without ascending the little slope leading to the rookery, and stayed there all day.

With the few exceptions I have noted above, all the birds that had arrived so far either were much fatigued, or else they realized that they had come a little too soon and were waiting for some psychological moment to arrive, for they were all strangely quiet and inactive.

FIG. 10. IN THE FOREGROUND A MATED PAIR HAVE BEGUN TO BUILD. BEHIND AND TO THE RIGHT TWO UNMATED HENS LIE IN THEIR SCOOPS

On October 18 the weather cleared and a fair number of penguins started to build their nests. The great majority however, apparently resting, still sat about. Those that built took their stones from old nests, as at present so many of these lay unoccupied. They made quite large nests, some inches high at the sides, with a comfortable hollow in the middle to sit in. The stone carrying (Fig. 20) was done by the male birds, the hens keeping continual guard over the nest, as otherwise the pair would have been robbed of the fruits of their labours as fast as they were acquired.

As I strolled through the rookery, most of the birds took little or no notice of me. Some, however, swore at me very savagely, and one infuriated penguin rushed at me from a distance of some ten yards, seizing the leg of my windproof trousers. In the morning quite a large number lay down on the sea-ice, a few yards short of the rookery, content apparently to have got so far. They lay there all day, motionless on their breasts, with their chins outstretched on the snow.

By the evening of October 18 most of the penguins had gathered in little groups on the nest-covered eminences, but there was at that time ample room for all, there being only about three or four thousand arrived. Although there were several open water holes against bergs frozen into the sea-ice some half mile or so away, not a single bird attempted to get food.

At 6 P.M. the whole rookery appeared to sleep, and the ceaseless chattering of the past hours gave place to a dead and impressive silence, though here and there an industrious little bird might be seen busily fetching stone after stone to his nest.

At that date it was deeply dusk at midnight, though the sun was very quickly rising in altitude, and continuous daylight would soon overtake us.

FIG. 11. THE ROOKERY BEGINNING TO FILL UP

By the morning of October 19 there had been a good many more arrivals, but the rookery was not yet more than one-twentieth part full. All the birds were fasting absolutely. Nest building was now in full swing, and the whole place waking up to activity. Most of the pebbles for the new nests were being taken from old nests, but a great deal of robbery went on nevertheless. Depredators when caught were driven furiously away, and occasionally chased for some distance, and it was curious to see the difference in the appearance between the fleeing thief and his pursuer. As the former raced and ducked about among the nests, doubling on his tracks, and trying by every means to get lost in the crowd and so rid himself of his pursuer, his feathers lay close back on his skin, giving him a sleek look which made him appear half the size of the irate nest-holder who sought to catch him, with feathers ruffled in indignation. This at first led me to think that the hens were larger than the cocks, as it was generally the hen who was at home, and the cock who was after the stones, but later I found that sex makes absolutely no difference in the size of the birds, or indeed in their appearance at all, as seen by the human eye. After mating, their behaviour as well as various outward signs serve to distinguish male from female. Besides this certain differences in their habits, which I will describe in another place, are to be noted.

The consciousness of guilt, however, always makes a penguin smooth his feathers and look small, whilst indignation has the opposite effect. Often when observing a knoll crowded with nesting penguins, I have seen an apparently under-sized individual slipping quietly along among the nests, and always by his subsequent proceedings he has turned out to be a robber on the hunt for his neighbours' stones. The others, too, seemed to know it, and would have a peck at him as he passed them.

At last he would find a hen seated unwarily on her nest, slide up behind her, deftly and silently grab a stone, and run off triumphantly with it to his mate who was busily arranging her own home. Time after time he would return to the same spot, the poor depredated nest-holder being quite oblivious of the fact that the side of her nest which lay behind her was slowly but surely vanishing stone by stone.

Here could be seen how much individual character makes for success or failure in the efforts of the penguins to produce and rear their offspring. There are vigilant birds, always alert, who seem never to get robbed or molested in any way: these have big high nests, made with piles of stones. Others are unwary and get huffed as a result. There are a few even who, from weakness of character, actually allow stronger natured and more aggressive neighbours to rob them under their very eyes.

Fig. 12. "The Hens would keep up this Peck-pecking hour after Hour"

FIG. 13. AN AFFECTIONATE COUPLE

In speaking of the robbery which is such a feature of the rookery during nest building, special note must be made of the fact that violence is never under any circumstances resorted to by the thieves. When detected, these invariably beat a retreat, and offer not the least resistance to the drastic punishment they receive if they are caught by their indignant pursuers. The only disputes that ever take place over the question of property are on the rare occasions when a *bona-fide* misunderstanding arises over the possession of a nest. These must be very rare indeed, as only on one occasion have I seen such a quarrel take place. The original nesting sites being, as I will show, chosen by the hens, it is the lady, in every case, who is the cause of the battle, and when she is won her scoop goes with her to the victor.

As I grew to know these birds from continued observation, it was surprising and interesting to note how much they differed in character, though the weaker-minded who would actually allow themselves to be robbed, were few and far between, as might be expected. Few, if any, of these ever could succeed in hatching their young and winning them through to the feathered stage.

When starting to make her nest, the usual procedure is for the hen to squat on the ground for some time, probably to thaw it, then working with her claws to scratch away at the material beneath her, shooting out the rubble behind her. As she does this she shifts her position in a circular direction until she has scraped out a round hollow. Then the cock brings stones, performing journey after journey, returning each time with one pebble in his beak which he deposits in front of the hen who places it in position.

Sometimes the hollow is lined with a neat pavement of stones placed side by side, one layer deep, on which the hen squats, afterwards building up the sides around her. At other times the scoop would be filled up indiscriminately by a heap of pebbles on which the hen then sat, working herself down into a hollow in the middle.

Individuals differ, not only in their building methods, but also in the size of the stones they select. Side by side may be seen a nest composed wholly of very big stones, so large that it is a matter for wonder how the birds can carry them, and another nest of quite small stones. (Fig. 14.)

Different couples seem to vary much in character or mood. Some can be seen quarrelling violently, whilst others appear most affectionate, and the tender politeness of some of these latter toward one another is very pretty to see. (Fig. 13.)

FIG. 14. "SIDE BY SIDE ... NESTS OF VERY BIG STONES AND NESTS OF VERY SMALL STONES"

I may here mention that the temperatures were rising considerably by October 19, ranging about zero F.

During October 20 the stream of arrivals was incessant. Some mingled at once with the crowd, others lay in batches on the sea-ice a few yards short of the rookery, content to have got so far, and evidently feeling the need for rest after their long journey from the pack. The greater part of this journey was doubtless performed by swimming, as they crossed open water, but I think that much of it must have been done on foot over many miles of sea-ice, to account for the fatigue of many of them.

Their swimming I will describe later. On the ice they have two modes of progression. The first is simple walking. Their legs being very short, their stride amounts at most to four inches. Their rate of stepping averages about one hundred and twenty steps per minute when on the march.

Their second mode of progression is "tobogganing." When wearied by walking or when the surface is particularly suitable, they fall forward on to their white breasts, smooth and shimmering with a beautiful metallic lustre in the sunlight, and push themselves along by alternate powerful little strokes of their legs behind them.

When quietly on the march, both walking and tobogganing produce the same rate of progression, so that the string of arriving birds, tailing out in a long line as far as the horizon, appears as a well-ordered procession. I walked out a mile or so along this line, standing for some time watching it tail past me and taking the photographs with which I have illustrated the scene. Most of the little creatures seemed much out of breath, their wheezy respiration being distinctly heard.

First would pass a string of them walking, then a dozen or so tobogganing. (Fig. 15.) Suddenly those that walked would flop on to their breasts and start tobogganing, and conversely strings of tobogganers would as suddenly pop up on to their feet and start walking. In this way they relieved the monotony of their march, and gave periodical rest to different groups of muscles and nerve-centres.

The surface of the snow on the sea-ice varied continually, and over any very smooth patches the pedestrians almost invariably started to toboggan, whilst over "bad going" they all had perforce to walk.

Figs. 16, 17, 18 and 19 present some idea of the procession of these thousands on thousands of penguins as day after day they passed into the rookery.

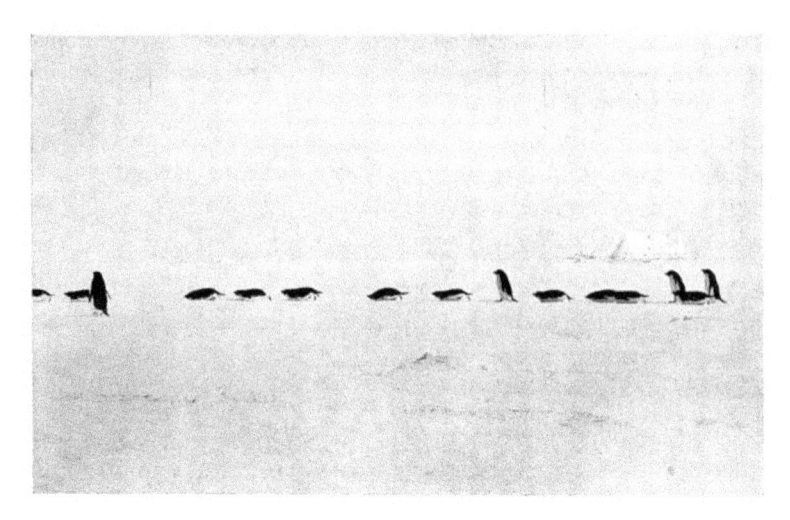

FIG. 15. ON THE MARCH TO THE ROOKERY OVER THE SEA-ICE. SOME ARE WALKING AND SOME "TOBOGGANING"

When tobogganing, turning to one side or the other is done with one or more strokes of the opposite flipper. When fleeing or chasing, both flippers as well as both feet are used in propulsion, and over most surfaces tobogganing is thus their fastest mode of progression, but when going at full tilt it is also the most exhausting, and after a short spurt in this way they invariably return to the walking position.

By October 20 many of the nests were complete, and the hens sat in them, though no eggs were to be seen yet. In the middle of one of the frozen lakes rose a little island, well suited for nesting except for the fact that later in the season, probably about the time when the young chicks were hatched, the lake would be thawed and the approach to the island only to be accomplished through about six inches or more of dirty water and ooze. Until then, however, the surface of the lake would remain frozen, and was at this time covered with snow.

Not a penguin attempted to build its nest on this island, though many passed it or walked over it in crossing the lake. How did they realize that later on they would get dirty every time they journeyed to or from the spot?

Not far from this island another mound rose from the lake, but this was connected with the "mainland" by a narrow neck of guano-covered pebbles. This mound was covered with nests, showing that the birds understood this place could always be reached over dry land. Surely this was well worth remarking.

There was a part of the ice-foot on the south side of the rookery where a track worn by many ascending penguins could be seen, leading from the sea-ice on to the beach. The place was steep and the ice slippery, and, in fact, the track led straight up a most difficult ascent. Not ten yards from this well-worn track a perfectly easy slope led up from the sea-ice to the rookery. The tracks in the freshly fallen snow showed that only one penguin had gone up this way. Presumably the first arrival in that place had taken the difficult path, and all subsequent arrivals blindly followed in his tracks, whilst only one had had the good luck or independence to choose the easier way.

On October 21 many thousands of penguins arrived from the northerly direction, and poured on to the beach in a continuous stream, the snaky line of arrivals extending unbroken across the sea-ice as far as the eye could see.

FIG. 16. PART OF THE LINE OF APPROACHING BIRDS, SEVERAL MILES IN LENGTH

A great many now started to climb the heights up the precipitous side of Cape Adare and to build their nests as far as the summit, a height of some 1000 feet, although there was still room for many thousand more down below. What could be their object, considering the wearisome journeys they would have to make to feed their young, it is impossible to say. It might be the result of the same spirit which made them spread out in little scattered groups over the rookery when only a few had arrived, and that they prefer wider room, only putting up with the greater crowding which ensues later as a necessary evil. There is, however, another explanation which I will discuss in another place.

At 9 P.M. it was getting dusk, and the rookery comparatively silent, although on some of the knolls two or three birds might be seen still busily working, toddling to and fro fetching stones. The other thousands lay at rest, their white breasts flat on the ground, and only their black beaks and heads visible as they lay with their chins stretched forward on the ground, whilst in place of the massed discord of clamour heard during the day, the separate voices of some of the busy ones were distinct. A fine powdering of snow was falling.

It would be difficult to estimate the number of penguins that poured into the rookery during the following day. There was no evidence that any pairing had taken place on or before the march, and the birds all had the appearance of being quite independent.

Far away from the beach the line had become thicker, and was no longer in single file, the progress of the birds being slow and steady, but when within half a mile or so from the beach, excitement seemed to take possession of them, and they would break into a run, hastening over the remaining distance, the line now being a thin one, with slight curves in it, each bird running, with wide gait, and outstretched flippers working away in unison with its little legs. In fact, the whole air of the line at this time was that of a school-treat arrived in sight of its playing-fields, and breaking into a run in its eagerness to get there.

FIG. 17. ARRIVING AT THE ROOKERY

Arrived at the rookery, and plunged suddenly amidst the din of that squalling, fighting, struggling crowd, the contrast with the dead silence and loneliness of the pack-ice they had so recently left, was as great a one as can well be imagined; yet once there, the birds seemed collected and at home. This was a matter of surprise to me then, but I remember now my own sensations on arriving home after my life in the Antarctic, and that I felt only slightly the sudden return to the bustle of civilization.

Our presence among them made little or no difference to the penguins. When we passed them closely they would bridle up and swear or even run at us and peck at our legs or batter them with their flippers, but unless their nesting operations were interfered with this attack was short-lived, and the next moment the birds would seem to forget our very existence. If I walked by the side of a long, nest-covered ridge, a low growl arose from every bird as I passed it, and the massed sound, gathering in front and dying away behind as I advanced, reminded me forcibly of the sound of the crowds on the towing-path at the 'Varsity boatrace as the crews pass up the river.

Walking actually among the nests, your temper is tried sorely, as every bird within reach has a peck at your legs, and occasionally a cock attacks you bravely, battering you with his little flippers in a manner ludicrous at first but aggravating after a time, as the operation is painful and severe enough to leave bruises behind it, and naturally this begins to pall. The courage of these little birds is most remarkable and admirable.

Our hut, being built on the rookery, could only be approached through crowds of penguins. Those that nested near us seemed quickly to become used to us and to take less notice of us than those farther off. One thing, however, terrified them pitiably. We had to fetch ice for our water from some stranded floes on the ice-foot, and this we did in a little sledge. As we hauled this rattling over the pebbly rookery it made a good deal of noise, and in its path nests were deserted, the occupants fleeing in the greatest confusion, a clear road being left for the sledge, whilst on either side a line of penguins was seen retreating in the utmost terror. After about a minute, they returned to their places and seemed to forget the incident, but we were very sorry to frighten them in this way, as we endeavoured to live at peace with them and to molest them as little as possible, and we feared that later on eggs might be spilt from the nests and broken. As time went on, those on the route of the sledge became accustomed even to this, and we were able to choose a course which cleared their nests.

FIG. 18. ARRIVING AT THE ROOKERY. IN THE BACKGROUND IS THE CLIFF UP WHICH MANY OF THE BIRDS CLIMB TO MAKE THEIR NESTS AT THE SUMMIT

Although squabbles and encounters had been frequent since their arrival in any numbers, it now became manifest that there were two very different types of battle; first, the ordinary quarrelling consequent on disputes over nests and the robbery of stones from these, and secondly, the battles between cocks who fought for the hens. These last were more earnest and severe, and were carried to a finish, whereas the first named rarely proceeded to extremes.

In regard to the mating of the birds, the following most interesting customs seemed to be prevalent.

The hen would establish herself on an old nest, or in some cases scoop out a hollow in the ground and sit in or by this, waiting for a mate to propose himself. (Fig. 26.) She would not attempt to build while she remained unmated. During the first week of the nesting season, when plenty of fresh arrivals were continually pouring into the rookery, she did not have long to wait as a rule. Later, when the rookery was getting filled up, and only a few birds remained unmated in that vast crowd of some three-quarters of a million, her chances were not so good.

For example, on November 16 on a knoll thickly populated by mated birds, many of which already had eggs, a hen was observed to have scooped a little hollow in the ground and to be sitting in this. Day after day she sat on looking thinner and sadder as time passed and making no attempt to build her nest. At last, on November 27, she had her reward, for I found that a cock had joined her, and she was busily building her nest in the little scoop she had made so long before, her husband steadily working away to provide her with the necessary pebbles. Her forlorn appearance of the past ten days had entirely given place to an air of occupation and happiness.

As time went on I became certain that invariably pairing took place after arrival at the rookery. On October 23 I went to the place where the stream of arrivals was coming up the beach, and presently followed a single bird, which I afterwards found to be a cock, to see what it was going to do. He threaded his way through nearly the whole length of the rookery by himself, avoiding the tenanted knolls where the nests were, by keeping to the emptier hollows. About every hundred yards or so he stopped, ruffled up his feathers, closed his eyes for a moment, then "smoothed himself out" and went on again, thus evidently struggling against desire for sleep after his journey. As he progressed he frequently poked his little head forward and from side to side, peering up at the knolls, evidently in search of something.

Fig. 19. Adélies arriving at the Rookery

FIG. 20. A COCK CARRYING A STONE TO HIS NEST

Arrived at length at the south end of the rookery, he appeared suddenly to make up his mind, and boldly ascending a knoll which was well tenanted and covered with nests, walked straight up to one of these on which a hen sat. There was a cock standing at her side, but my little friend either did not see him or wished to ignore him altogether. He stuck his beak into the frozen ground in front of the nest, lifted up his head and made as if to place an imaginary stone in front of the hen, a most obvious piece of dumb show. The hen took not the slightest notice nor did her mate.

My friend then turned and walked up to another nest, a yard or so off, where another cock and hen were. The cock flew at him immediately, and after a short fight, in which each used his flippers savagely, he was driven clean down the side of the knoll away from the nests, the victorious cock returning to his hen. The newcomer, with the persistence which characterises his kind, came straight back to the same nest and stood close by it, soon ruffling his feathers and evidently settling himself for a doze, but, I suppose, because he made no further overtures the others took no notice of him at all, as, overcome by sheer weariness, he went to sleep and remained so until I was too cold to await further developments. On my way back to our hut I followed another cock for about thirty yards, when he walked up to another couple at a nest and gave battle to the cock. He, too, was driven off after a short and decisive fight. Soon there were many cocks on the war-path. Little knots of them were to be seen about the rookery, the lust of battle in them, watching and fighting each other with desperate jealousy, and the later the season advanced the more "bersac" they became.

A typical scene I find described in my notes for October 25 when I was out with my camera, and I mention it as a type of the hundreds that were proceeding simultaneously over the whole rookery, and also because I was able to photograph different stages of the proceedings as follows:

Fig. 22 shows a group of three cocks engaged in bitter rivalry round a hen who is cowering in her scoop in which she had been waiting as is their custom. She appeared to be bewildered and agitated by the desperate behaviour of the cocks.

FIG. 21. SEVERAL INTERESTING THINGS ARE TAKING PLACE HERE

On Fig. 23 a further development is depicted, and two of the cocks are seen to be squaring up for battle. Close behind and to the right of them are seen (from left to right) the hen and the third cock, who are watching to see the result of the contest, and another hen cowering for protection against a cock with whom she has become established.

Fig. 24 shows the two combatants hard at it, using their weight as they lean their breasts against one another, and rain in the blows with their powerful flippers.

Fig. 25 shows the end of the fight, the victor having rushed the vanquished cock before him out of the crowd and on to a patch of snow on which, as he was too brave to turn and run, he knocked him down and gave him a terrible hammering.

When his conqueror left him at length, he lay for some two minutes or so on the ground, his heaving breast alone showing that he was alive, so completely exhausted was he, but recovering himself at length he arose and crawled away, a damaged flipper hanging limply by his side, and he took no further part in the proceedings. The victorious bird rushed back up the side of the knoll, and immediately fought the remaining cock, who had not moved from his original position, putting him to flight, and chasing him in and out of the crowd, the fugitive doubling and twisting amongst it in a frantic endeavour to get away, and I quickly lost sight of them.

Scenes of this kind became so common all over the rookery, that the roar of battle and thuds of blows could be heard continuously, and of the hundreds of such fights, all plainly had their cause in rivalry for the hens.

When starting to fight, the cocks sometimes peck at each other with their beaks, but always they very soon start to use their flippers, standing up to one another and raining in the blows with such rapidity as to make a sound which, in the words of Dr. Wilson, resembles that of a boy running and dragging his hoop-stick along an iron paling. Soon they start "in-fighting," in which position one bird fights right-handed, the other left-handed; that is to say, one leans his left breast against his opponent, swinging in his blows with his right flippers, the other presenting his right breast and using his left flipper. My photographs of cocks fighting all show this plainly. It is interesting to note that these birds, though fighting with one flipper only, are ambidextrous. Whilst battering one another with might and main they use their weight at the same time, and as one outlasts the other, he drives his vanquished opponent before him over the ground, as a trained boxing man, when "in-fighting" drives his exhausted opponent round the ring.

Fig. 22. Three Cocks in Rivalry

FIG. 23. TWO OF THE COCKS SQUARING UP FOR BATTLE

Desperate as these encounters are, I don't think one penguin ever kills another. In many cases blood is drawn. I saw one with an eye put out, and that side of its beak (the right side) clotted with blood, whilst the crimson print of a blood-stained flipper across a white breast was no uncommon sight.

Hard as they can hit with their flippers, however, they are also well protected by their feathers, and being marvellously tough and enduring the end of a hard fight merely finds the vanquished bird prostrate with exhaustion and with most of the breath beaten out of his little body. The victor is invariably satisfied with this, and does not seek to dispatch him with his beak.

It was very usual to see a little group of cocks gathered together in the middle of one of the knolls squabbling noisily. Sometimes half a dozen would be lifting their raucous voices at one particular bird, then they would separate into pairs, squaring up to one another and emphasizing their remarks from time to time by a few quick blows from their flippers. It seemed that each was indignant with the others for coming and spoiling his chances with a coveted hen, and trying to get them to depart before he went to her.

It was useless for either to attempt overtures whilst the others were there, for the instant he did so, he would be set upon and a desperate fight begin. Usually, as in the case I described above, one of the little crowd would suddenly "see red" and sail into an opponent with desperate energy, invariably driving him in the first rush down the side of the knoll to the open space surrounding it, where the fight would be fought out, the victor returning to the others, until by his prowess and force of character, he would rid himself of them all. Then came his overtures to the hen. He would, as a rule, pick up a stone and lay it in front of her if she were sitting in her "scoop," or if she were standing by it he might himself squat in it. She might take to him kindly, or, as often happened, peck him furiously. To this he would submit tamely, hunching up his feathers and shutting his eyes while she pecked him cruelly. Generally after a little of this she would become appeased. He would rise to his feet, and in the prettiest manner edge up to her, gracefully arch his neck, and with soft guttural sounds pacify her and make love to her.

Fig. 24. "Hard at it"

FIG. 25. THE END OF THE BATTLE

Both perhaps would then assume the "ecstatic" attitude, rocking their necks from side to side as they faced one another (Fig. 26), and after this a perfect understanding would seem to grow up between them, and the solemn compact was made.

It is difficult to convey in words the daintiness of this pretty little scene. I saw it enacted many dozens of times, and it was wonderful to watch one of these hardy little cocks pacifying a fractious hen by the perfect grace of his manners.

Fig. 21 is particularly instructive. In the centre of the picture a group of cocks are quarrelling, and on the left-hand side three unmated hens can be seen sitting in their scoops, whilst two of them (the two in front) are receiving overtures from two of the cocks who are making the most of their time whilst the others are fighting. On the right-hand side another cock is seen proposing himself to a fourth hen who seems to be meeting his overtures with the usual show of reluctance.

Although for the later arrivals a good deal of fighting was necessary before a mate could be secured, it seemed that some got the matter fixed up without any difficulty at all, especially during the earlier days when only a few birds were scattered widely over the rookery. Later, the cocks seemed to watch one another jealously, and to hunt in little batches in consequence. (Figs. 27, 28, and 29.)

From the particulars I have just given it is also evident that a wife and home once obtained could only be kept by dint of further battling and constant vigilance during the first stages of domesticity, when thousands of lusty cocks were pouring into the rookery, and it was not unusual to see a strange cock paying court to a mated hen in the absence of her husband until he returned to drive away the interloper, but I do not think that this ever occurred after the eggs had come and the regular family life begun, couples after this being perfectly faithful to one another.

The instance I have given of a newly arrived cock by dumb show pretending to take a stone and place it before a mated hen, is typical of the sort of first overture one sees, though more frequently an actual stone was tendered. While on this subject I had better mention a most interesting thing which occurred to one of my companions. One day as he was sitting quietly on some shingle near the ice-foot, a penguin approached him, and after eyeing him for a little, walked right up to him and nibbled gently at one of the legs of his wind-proof trousers. Then it walked away, picked up a pebble, and came back with it, dropping it on the ground by his side. The only explanation of this occurrence seems to be that the tendering of the stone was meant as an overture of friendship.

FIG. 26. THE PROPOSAL. (NOTE THE HEN IN HER SCOOP)

On October 26 there was no abatement in the stream of arrivals. The cockfighting continued, and many of them, temporarily disabled, were to be seen moping about the rookery, smeared with blood and guano. Often a hen would join in when two cocks were fighting, occasionally going first for one and then the other, but I never to my knowledge saw a cock retaliate on a hen.

Once I saw two cocks fighting, and a hen taking the part of one of the cocks, the pair of them gave the other a fearful hammering, the hen using her bill savagely as well as her flippers. Completely knocked out and gasping for breath he got away at last, only to meet another cock who fought him and easily beat him. When this one had gone a third came, and the poor victim with a courage truly noble was squaring himself up with his last spark of energy, when I interfered and drove away his enemy.

The nests on most of the knolls soon became so crowded that their occupants, by stretching out their necks, could reach their neighbours without getting up. As every hen appeared to hate her neighbour they would peck-peck at one another hour after hour, in the manner seen in my photograph,(3) till their mouths and heads became terribly sore. Occasionally they would desist, shake their heads apparently from pain, then at it again.

In various places through the course of these pages, reference is made to the "ecstatic" attitude of the penguins. This antic is gone through by both sexes and at various times, though much more frequently during the actual breeding season. The bird rears its body upward and stretching up its neck in a perpendicular line, discharges a volley of guttural sounds straight at the unresponding heavens. At the same time the clonic movements of its syrinx or "sound box" distinctly can be seen going on in its throat. Why it does this I have never been able to make out, but it appears to be thrown into this ecstasy when it is pleased; in fact, the zoologist of the "Pourquoi Pas" expedition termed it the "Chant de satisfaction." I suppose it may be likened to the crowing of a cock or the braying of an ass. When one bird of a pair starts to perform in this way, the other usually starts at once to pacify it. Very many times I saw this scene enacted when nesting was in progress. The two might be squatting by the nest when one would arise to assume the "ecstatic" attitude and make the guttural sounds in its syrinx. Immediately the other would get close up to it and make the following noise in a soft soothing tone:

A-ah

Always and immediately this caused the musician to subside and settle itself down again.

Fig. 27. Cocks fighting for Hens

FIG. 28. COCKS FIGHTING FOR HENS

The King penguin at the Zoological Gardens, whose sex is unknown, throws itself into the ecstatic attitude and sings a sort of song when its keeper strokes its neck. The blackfooted penguins never do it, though they breed several times a year. Figs. 26 and 32 show Adélies in ecstatic attitude.

To-day about a dozen skua gulls (*Megalestris Makormiki*) appeared for the first time. They did not start to nest, but sat on the sea-ice with a group of penguins, in apparent amity. A few occasionally flew about over the rookery.

On October 27 though the stream of arrivals continued there were wide gaps in it. It appeared to be thinning. For an hour in the forenoon it stopped altogether, and at the end of this time a storm of wind from the south struck us and continued for another hour with thick drift. Probably clear of Cape Adare the wind had been blowing before it reached us, and had stopped the birds' progress across the ice.

During the storm the rookery was completely silenced, most of the birds lying with their heads to the wind. A good many skuas arrived that day. Some chips of white, glistening quartz had been thrown down by our hut door recently, and later I found two of these chips in a nest about thirty yards away, showing up brightly against the black basalt of which all the pebbles on the rookery were composed.

As a rule the penguins were careful to select rounded stones for their nests, but these fragments of quartz were jagged and uncomfortable, and most unsuitable for nest building. Thus it was evidently the brightness of the stones which attracted them. Whilst I looked on, the owners of the pieces of quartz were wrangling with their neighbours, and a penguin in a nest behind shot out its beak and stole one of the pieces, placing it in its own nest. I had brought Campbell out to show him the pieces of quartz, and he witnessed the last incident with me.

FIG. 29. COCKS FIGHTING FOR HENS

Fig. 30. Penguin on Nest

I may here mention an experiment I tried some days later. I painted some pebbles a bright red and had others covered with bright green cotton material as I had no other coloured paint. Mixing a handful of these coloured stones together I placed them in a little heap amongst natural black ones near a nest-covered knoll. Returning in a few hours I found nearly all the red stones and one or two of the green ones gone, and later found them in nests. Later still, all the red ones had disappeared, and last of all the green ones. I traced nearly all these to nests, and found a few days later that, like the pieces of white quartz, they were being stolen from nest to nest and thus slowly being distributed in different directions. At other times I saw pieces of tin, pieces of glass, half a stick of chocolate, and the head of a bright metal teaspoon in different nests near our hut, the articles evidently having been taken from our scrap-heap. Thus it is evident that penguins like bright colours and prefer red to green, as instanced by the selection of the coloured pebbles. I am sorry that I did not carry these colour tests further.

During October 29 the stream of arrivals was undiminished, but the next day it slackened considerably, and during the next two days stopped altogether, all the rising ground of the rookery now being literally crammed full with nests, several thousands of them being scattered up the slopes of Cape Adare to a height of a thousand feet.

Fig. 31. Showing the Position of the Two Eggs

Fig. 32. An Adélie in "Ecstatic" Attitude

PART II

DOMESTIC LIFE OF THE ADÉLIE PENGUIN

Laying and incubation of the eggs : The Adélies' habits in the water : Their games : Care of the young : The later development of their social system.

ON November 3 several eggs were found, and on the 4th these were beginning to be plentiful in places, though many of the colonies had not yet started to lay.

Let me here call attention to the fact that up to now not a single bird out of all those thousands had left the rookery once it had entered it. Consequently not a single bird had taken food of any description during all the most strenuous part of the breeding season, and as they did not start to feed till November 8 thousands had to my knowledge fasted for no fewer than twenty-seven days. Now of all the days of the year these twenty-seven are certainly the most trying during the life of the Adélie.

With the exception, in some cases, of a few hours immediately after arrival (and I believe the later arrivals could not afford themselves even this short respite) constant vigilance had been maintained; battle after battle had been fought; some had been nearly killed in savage encounters, recovered, fought again and again with varying fortune. They had mated at last, built their nests, procreated their species, and, in short, met the severest trials that Nature can inflict upon mind and body, and at the end of it, though in many cases blood-stained and in all caked and bedraggled with mire, they were as active and as brave as ever.

When one egg had been laid the hen still sat on the nest. The egg had to be continually warmed, and as the temperature was well below freezing-point, exposure would mean the death of the embryo.

In order to determine the period between the laying of the two eggs, I numbered seven nests with wooden pegs, writing on the pegs the date on which each egg was laid. The result obtained is shown on page 53.

The average interval in the four cases where two eggs were laid being 3·5 days.

FIG. 33. FLOODS

No. 7 nest was that of the hen which I mentioned as having waited for so long for a mate, and the lateness of the date on which the first egg appeared may have resulted in there being no other.

	Date of appearance of first egg	Date of appearance of second egg	Interval
No. 1 nest	Nov. 14	—	Only 1 laid
No. 2 nest	Nov. 13	Nov. 16	3 days
No. 3 nest	Nov. 14	Nov. 17	3 days
No. 4 nest			
No. 5 nest	Nov. 12	Nov. 16	4 days
No. 6 nest	Nov. 8	Nov. 12	4 days
No. 7 nest	Nov. 24	—	Only 1 laid

The only notes I have on the incubation period are that the first chick appeared in No. 5 nest on December 19 (incubation period thirty-seven days) and in No. 7 nest on December 28 (incubation period thirty-four days).

The skuas had increased considerably in numbers by November 4, and frequently came to the scrap-heap outside our hut. Here were many frozen carcasses of penguins which we had thrown there after the breasts had been removed for food during the past winter. The skuas picked the bones quite clean of flesh, so that the skeletons lay white under the skins, and it was remarkable to what distances they sometimes carried the carcasses, which weighed considerably more than the skuas themselves. I found some of these bodies over five hundred yards away.

A perpetual feud was carried on between the penguins and the skuas. The latter birds come to the south in the summer, and make their nests close to, and in some cases actually among, those of the penguins, and during the breeding time live almost entirely on the eggs, and later, on the chicks. They never attack the adult penguins, who run at them and drive them away when they light within reach, but as the skuas can take to the wing and the penguins cannot, no pursuit is possible.

Fig. 34. Flooded

Fig. 35. A Nest with Stones of Mixed Sizes

The skuas fly about over the rookery, keeping only a few yards from the ground, and should one of them see a nest vacated and the eggs exposed, if only for a few seconds, it swoops at this, and with scarcely a pause in its flight, picks up an egg in its beak and carries it to an open space on the ground, there to devour the contents. Here then was another need for constant vigilance, and so daring did the skuas become, that often when a penguin sat on a nest carelessly, so as to leave one of the eggs protruding from under it, a lightning dash from a skua would result in the egg being borne triumphantly away.

The bitterness of the penguins' hatred of the skuas was well shown in the neighbourhood of our scrap-heap. None of the food thrown out on to this heap was of the least use to the penguins, but we noticed after a time that almost always there were one or more penguins there, keeping guard against the skuas, and doing their utmost to prevent them from getting the food, and never allowing them to light on the heap for more than a few seconds at a time. In fact, a constant feature of this heap was the sentry penguin, darting hither and thither, aiming savage pecks at the skuas, which would then rise a yard or two into the air out of reach, the penguin squalling in its anger at being unable to follow its enemy. At this juncture the penguin would imitate the flying motion with its flappers, seeming instinctively to attempt to mount into the air, as its remote ancestors doubtlessly did, before their wings had adapted themselves solely to swimming.

Close to the scrap-heap there was a large knoll crowded with penguins' nests, and it was this knoll that provided the sentries. Very rarely did one of these leave the heap until another came to relieve it as long as there were skuas about, but when the skuas went the penguins left it too. When the skuas returned, however, and without the lapse of a few seconds, a penguin would be seen to detach itself from the knoll and run to guard the heap. That some primitive understanding on this matter existed among the penguins seems to me probable, because whilst there were generally one or two guarding the heap, there was never a crowd, the rest of the knoll seeming quite satisfied as long as one of their number remained on guard.

In describing the Cape Adare rookery I mentioned the fact that the pebbles entering into the formation of the beach are basaltic, and therefore of a dead black shade. The result of this is that as the sun's altitude increases, heat is absorbed readily by the black rock, through that clear atmosphere, and the snow upon it rapidly melts.

FIG. 36. "HOUR AFTER HOUR, DURING THE WHOLE DAY, THEY FOUGHT AGAIN AND AGAIN"

For a long time the penguins at their nests had satisfied their thirst by eating the snow near them, but as this disappeared, they suffered greatly, as was made evident by the way they lay with beaks open and tongues exposed between them. (Fig. 30.) As time went on the cocks started to make short journeys to the drifts which still remained in order to quench their thirst, but the hens stuck manfully, or rather "henfully" to their posts, though some of them seemed much distressed. Later, those cocks which had nested in the centre of the rookery had quite long journeys to make in order to find drifts, a very popular resort being that which had formed in the lee of our hut, and all day streams of them came here to gobble snow. Once a cock was seen to take a lump of snow in his beak and carry it to his mate on the nest, who ate it.

Mr. Priestley tells me that when he was at Cape Royds in 1908 he saw cocks taking snow to hens on their nests. This procedure would seem to be different to the parental instinct which governs the feeding of the young, and it seemed to show that the cock realized that the hen must be thirsty and in need of the snow, and kept this fact in mind when he was away from her. Another point to note is that the occurrence was a very rare and, in fact, exceptional one.

When conditions arose which were new to their experience the penguins seemed utterly unable to grasp them.

As an example of this, we had rigged a guide rope from our hut to the meteorological screen, about fifty yards away, to guide us during blizzards. This rope, which was supported by poles driven into the ground, sagged in one place till it nearly touched the ground. At frequent intervals, penguins on their way past the hut were brought to a standstill by running their breasts into this sagged rope, and each bird as it was caught invariably went through the same ridiculous procedure. First it would push hard against the rope, then finding this of no avail, back a few steps, walk up to it again and have another push, repeating the process several times. After this, instead of going a few feet further along where it could easily walk under the rope, in ninety per cent. of cases it would turn, and by a wide detour walk right round the hut the other way, evidently convinced that some unknown obstacle completely barred its passage on that side. This spectacle was a continual source of amusement to us as it went on all day and every day for some time.

FIG. 37. A NEST ON A ROCK

FIG. 38. "ONE AFTER ANOTHER, THE REST OF THE PARTY FOLLOWED HIM"

As penguins' eggs are very good to eat and a great luxury, as well as being beneficial to men living under Antarctic conditions, we collected a large number, which we stowed away to freeze. To collect these eggs we used to set off, carrying a bucket, and walk through the knolls. As we picked our way, carefully placing our feet in the narrow spaces between the nests, we were savagely pecked about the legs, as in most positions at least, these birds could reach us without even leaving the nest, whilst very often the mates standing near them would sail in at us, raining in blows with their flippers with the rapidity of a maxim gun.

To search for eggs it was necessary to lift up the occupant of each nest and look beneath her. If she were tackled from front or flank this was a painful and difficult business, as she drove at the intruder's hands with powerful strokes of her sharp beak, but we found that the best way to set about the matter was to dangle a fur mit in front of her with one hand, and when she seized this quickly slip the other behind her, lifting her nether regions from the nest, and at the same time pushing her gently forward. Immediately she would drop the fur mit, and sticking her beak into the ground push herself backward with a determined effort to stay on the nest. So long as the pressure from behind was kept up she would keep her beak firmly fixed in the ground, and could be robbed at will.

The egg abstracted, she was then left in peace, on which she would rise to her feet, look under her for the egg and, finding that it was gone, ruffle her feathers, and, trembling with indignation, look round for the robber, seemingly quite unable to realize that we were the guilty ones. This is typical of the Adélie's attitude towards us. We are beyond their comprehension, and fear of us, anger at us, curiosity over us, although frequently shown, are displayed only for a fleeting moment. In a few minutes she might forget about the incident altogether and quietly resume her position on the empty nest, but very often she would violently attack any other bird who might happen to be standing near, and thus as we filled our buckets we left a line of altercation in our wake. This, however, was not long lived, and affairs soon settled down to their normal state, and I believe that in about one minute the affair was completely forgotten. The penguin, indeed, is in its nature the embodiment of all that man should be when he explores the Antarctic regions, ever acting on the principle that it is of no use to worry over spilt milk.

FIG. 39. A JOY RIDE

The comparative size of the penguin's egg is shown in some of my photographs. Ninety-six eggs averaged 4·56 ounces apiece. They vary in size from about 6·45 cm. to 7·2 cm. in length, and from 5·0 cm. to 5·5 cm. in breadth, on an average. Both ends are nearly equally rounded, and of a white chalky texture without, and green within. This green colour is plainly shown by transmitted light.

When the two have been laid the sitting bird places them one in front of the other. The rearmost egg is tucked up on the outspread feet, the foremost lies on the ground, and is covered by the belly of the bird as it lies forward upon it. (Fig. 31.) By many of the birds a strong inclination to burrow was displayed, and they seemed very fond of delving in the soft shingle ledges that were to be found on some parts of the beach. They did this ostensibly to get small stones for their nests, but certainly burrowed deeper than they need have done, and occasionally squatted for some time in little caves that they made in this way. I noticed the same thing in the drifts when they went to eat snow, and thought at times that they were going to make underground nests, but they never did so, though some of the little shingle caves would have made ideal nesting sites.

By November 7, though many nests were still without eggs, a large number now contained two, and their owners started, turn and turn about, to go to the open water leads about a third of a mile distant to feed, and as a result of this a change began gradually to come over the face of the rookery. Hitherto the whole ground in the neighbourhood of the nests had been stained a bright green. This was due to the fasting birds continually dropping their watery, bile-stained excreta upon it. (The gall of penguins is bright green.) These excreta practically contained no solid matter excepting epithelial cells and salts.

The nests themselves are never fouled, the excreta being squirted clear of them for a distance of a foot or more, so that each nest has the appearance of a flower with bright green petals radiating from its centre. Some of the photographs show this well, especially Fig. 30. Even when the chicks have come and are being sat upon by the parents, this still holds good, because they lie with their heads under the old bird's belly and their hindquarters just presenting themselves, so that they may add their little decorative offerings, petal by petal! Now that the birds were going to feed, the watery-green stains upon the ground gave place to the characteristic bright brick-red guano, resulting from their feeding on the shrimp-like euphausia in the sea; and the colour of the whole rookery was changed in a few days, though this was first noticeable, of course, in the region of those knolls which had been occupied first, and which were now settled down to the peaceable and regular family life which was to last until the chicks had grown.

FIG. 40. A KNOT OF PENGUINS ON THE ICE-FOOT

As this family life became established, law and order reigned to some extent, and there was a distinct tendency to preserve it, noticeably on those knolls which had so settled down, and I think the following most surprising incident bears evidence of what I have said. I quote word for word from my notes on November 24, 1911:

"This afternoon I saw two cocks (probably) engaged in a very fierce fight, which lasted a good three minutes. They were fighting with flippers and bills, one of them being particularly clever with the latter, frequently seizing and holding his opponent just behind the right eye whilst he battered him with his flippers.

"After a couple of minutes, during which each had the other down on the ground several times, three or four other penguins ran up and apparently tried to stop the fight. This is the only construction I can put on their behaviour, as time after time they kept running in when the two combatants clinched, pushing their breasts in between them, but making no attempt to fight themselves, whilst their more collected appearance and smooth feathers were in marked contrast to the angry attitudes of the combatants.

"The fight, which had started on the outskirts of a knoll crowded with nests, soon edged away to the space outside, and it was here that I (and Campbell, who was with me) saw the other penguins try to stop it. The last minute was a very fierce and vindictive 'mill,' both fighting with all their might, and ended in one of them trying to toboggan away from his opponent; but he was too exhausted to get any pace on, so that just as he got into the crowd again he was caught, and both fought for a few seconds more, when the apparent victor suddenly stopped and ran away. The other picked himself up and made his way rapidly among the nests, evidently searching for one in particular.

"Following him, I saw him run up to a nest near the place where the fight had begun. There was a solitary penguin waiting by this nest, which was evidently new and not yet completed, and without eggs. The cock I had followed, ruffled and battered with battle, ran up to the waiting bird, and the usual side-to-side chatter in the ecstatic attitude began and continued for half a minute, after which each became calmer, and I left them apparently reconciled and arranging stones in the nest.

"This incident was after the usual nature of a dispute between two mates for a hen, but the pacific interference of the other birds was quite new to my experience. That it was pacific I am quite convinced, and Campbell agreed with me that there was no doubt of it. All the nests round about had eggs under incubation, and the pair in question must have been newcomers."

FIG. 41. AN ADÉLIE LEAPING FROM THE WATER

On returning home I was glad to find that Mr. Bernacchi, who landed at Cape Adare with the "Southern Cross" expedition, says in his account (p. 131) that he also saw penguins interfering and trying to stop others from fighting.

Owing to our having several snowfalls without wind, and to the action of the sun on the black rock, which I have mentioned already, the rookery became a mass of slush in many places, and in some of the lower-lying parts actually flooded. In some of these low-lying situations penguins had unwarily made their nests, and there was one particular little colony near our hut which was threatened with total extinction from the accumulation of thaw water. As this trickled down from the higher ground around them, the occupants of the flooded ground exerted all their energies to avert this calamity, and from each nest one of its tenants could be seen making journey after journey for pebbles, which it brought to the one sitting on the nest, who placed stone after stone in position, so that as the water rose the little castle grew higher and higher and kept the eggs dry. One nest in particular I noticed which was as yet a foot or so clear of the water and on dry ground; but whilst the hen sat on this, the cock was working most energetically in anticipation of what was going to happen, and for hours journeyed to and from the nest, each time wading across the little lake to the other side, where he was getting the stones.

This scene, which I photographed, is depicted on Fig. 33. In the right-hand corner of the picture the cock is seen in the act of delivering another stone to the hen who is waiting to receive it, whilst some of the nests are actually surrounded by water. Fig. 34 shows another nest, rising like a little island from a thaw pool, the eggs being only just above water.

FIG. 42. ADÉLIE LEAPING FROM THE WATER (THIS BIRD JUMPED 4 FEET HIGH AND 10 FEET LONG)

Some time ago I mentioned that there were penguins of weak individuality who allowed others to rob them of their stones, and this was in some cases very noticeable on the flooded ground, and there were one or two nests here which had been almost entirely removed by thieving neighbours.

To quote again from my notes.

"November 10. This evening I saw a hen penguin trying to sit on a nest with two eggs. The nest had no stones, and was scooped deeply in the ground in a slush of melting snow, so that the eggs were nearly covered with water. The poor hen stood in the water and kept trying to squat down on the eggs, but each time she did so, sat in the water and had to get up again. She was shivering with cold and all bedraggled.

"I took the two eggs out of the nest, and Browning and I collected a heap of stones (partly from her richer neighbours!) and built the nest well up above the water. Then I replaced the eggs, and the hen at once gladly sat on them, put them in position, and was busily engaged in arranging the new stones round her when we left."

One day, when the season was well advanced, I saw a violent altercation taking place between two penguins, one of which was in possession of a nest in a somewhat isolated position. The other evidently was doing his utmost to capture the nest, as whenever he got the other off, he stood on it. There were scarcely any stones in the nest, which contained one egg. I think from the way they fought that both were cocks.

For two reasons I make special mention of the occurrence, first, because of all the fights I ever saw this was the longest and most relentless, and, secondly, because the nest being in such an isolated position it seemed curious that there could be any mistake about its ownership. Such, however, seemed to be the case, and hour after hour, during the whole day, they fought again and again.

After each bout of a few minutes both birds became so exhausted that they sank panting to the ground, evidently suffering from thirst and at the limit of their endurance. Sometimes one captured the nest, sometimes the other, but after several hours of this, one of them began to show signs of outlasting the other, and kept possession. For long after this, however, the other returned repeatedly to the attack.

FIG. 43. JUMPING FROM THE WATER ON TO SLIPPERY ICE

I fetched my camera and photographed the birds as they fought (Fig. 36). As time went on, the weaker bird took longer and longer intervals to recover between his attacks, lying on his breast, with his head on the snow and eyes half closed, so that I thought he was going to die. Each time he got to his feet and staggered at his enemy, the latter rose from the nest and met him, only to drive him back again. When I saw them at about 10 P.M. (it was perpetual daylight now) both were lying down, the victor on the nest, the vanquished about five yards off. The next day one bird remained on the nest and the other had gone, and I do not know what happened to him.

In the course of a walk through the rookery considerable diversity in the choice of nesting sites was to be noticed. The general tendency is for the penguins to build their nests close together (within a foot or two of one another) on the tops of the rounded knolls, the lower levels being left untenanted.

The most thickly populated districts were to be found on the screes immediately below the cliffs. These screes having been formed in the first instance by the falling of fragments due to weathering of the cliff, their substance is still added to, little by little, as time goes on, and therefore many are killed annually by falling rocks, as is mentioned elsewhere, but weighing against this danger is the advantage the cliff offers as a shelter from the E.S.E. gales. The same applies to the nesting sites up the cliff, but I am convinced that only the love of climbing can account for the extraordinary positions chosen by some of the birds. Some of the nests are so difficult of access that their occupants, on their way to them, may be seen sliding backwards down the little glazed snow-slopes several times before they accomplish the ascent, whilst in other places they have to jump from one foothold to another along the almost perpendicular cliff.

Even up these heights a tendency to grouping is seen, though there are a fair number of individuals who, seeming to seek seclusion, make their nests at some distance from the others. I noticed this in some places along the shore, too, where solitary nests were to be seen on isolated patches of shingle.

When I visited Cape Royds in 1911 I found a couple nesting alone in a cove known as "Black Sand Beach," some half mile from the rookery there. Such isolation as this, however, is very unusual, and was quite a departure from the regular custom of the species.

FIG. 44. "WHEN THEY SUCCEEDED IN PUSHING ONE OF THEIR NUMBER OVER, ALL WOULD CRANE THEIR NECKS OVER THE EDGE"

In some places at Cape Adare, large rocks some two or three feet in height stood about the rookery. Whenever the summit of one of these was accessible, a pair built their nest upon it,(4) though how they managed to keep up there during the gales was a matter for wonder, but the proud possessors of the castle evidently had a delight in their lofty position. One nest had been made on an old packing-case left by the expedition which wintered there in 1894, and several nested among the weathering bones of the seals that had died on the beach.

Although the greatest care had been taken by nearly all in the choice of sites that would be on dry ground when the thaw came later in the season, yet a few hens had gone to the other extreme, and with greatest stupidity chosen their site right down in the hollows where they were absolutely certain to be flooded later on. These stupid ones are thus prevented from rearing their young, and so selection keeps the wiser for future generations, and eliminates the less intelligent from the community, though perhaps some of these learn by experience, and next year use more discrimination in choosing their nesting place.

Some of the colonies—in fact, most of them—were orderly and well arranged, and later in the season distinctly peaceful. Others, however, presented a less respectable appearance. There was one in particular, close to our hut, which could only be described as a slum of the meanest description. All through the season there was more fighting in this colony than anywhere else, and so remarkable was this, that we christened the locality "Casey's Court" and the name stuck for the rest of the year.

The nests had fewer stones than elsewhere, and were more untidily made, and when the eggs came, owing to the constant fighting that went on, most of them got spilt from the nests or broken, and very few chicks were hatched in consequence, the mortality among them also being so great that of the whole colony of some hundred nests, I do not think more than forty or fifty chicks at most reached maturity. The explanation of this state of things lay, I believe, in the fact that our hut and its curtilage deflected the stream of penguins on their way past the spot from the water to the back of the rookery, so that a constant stream of them passed through "Casey's Court," upsetting the tempers of the inhabitants so that they became disorderly. In addition to this, there was a fairly big thaw pool and much miry ground near by, so that the inhabitants were generally covered with mud and very disreputable to look at.

FIG. 45. DIVING FLAT INTO SHALLOW WATER

During the fasting season, as none of the penguins had entered the water, they all became very dirty and disreputable in appearance, as well may be imagined considering the life they led, but now that they went regularly to swim, they immediately got back their sleek and spotless state.

From the ice-foot to the open water, the half mile or so of sea-ice presented a lively scene as the thousands of birds passed to and fro over it, outward bound parties of dirty birds from the rookery passing the spruce bathers, homeward bound after their banquet and frolic in the sea. So interesting and instructive was it to watch the bathing parties, that we spent whole days in this way.

As I have said before, the couples took turn and turn about on the nest, one remaining to guard and incubate while the other went off to the water.

On leaving their nests, the birds made their way down the ice-foot on to the sea-ice. Here they would generally wait about and join up with others until enough had gathered together to make up a decent little party, which would then set off gaily for the water. They were now in the greatest possible spirits, chattering loudly and frolicking with one another, and playfully chasing each other about, occasionally indulging in a little friendly sparring with their flippers.

Arrived at length at the water's edge, almost always the same procedure was gone through. The object of every bird in the party seemed to be to get one of the others to enter the water first. They would crowd up to the very edge of the ice, dodging about and trying to push one another in. Sometimes those behind nearly would succeed in pushing the front rank in, who then would just recover themselves in time, and rushing round to the rear, endeavour to turn the tables on the others. Occasionally one actually would get pushed in, only to turn quickly under water and bound out again on to the ice like a cork shot out of a bottle. Then for some time they would chase one another about, seemingly bent on having a good game, each bird intent on finding any excuse from being the first in. Sometimes this would last a few minutes, sometimes for the better part of an hour, until suddenly the whole band would change its tactics, and one of the number start to run at full tilt along the edge of the ice, the rest following closely on his heels, until at last he would take a clean header into the water. One after another the rest of the party followed him (Fig. 38), all taking off exactly from the spot where he had entered, and following one another so quickly as to have the appearance of a lot of shot poured out of a bottle into the water. The accompanying photograph presents this last scene.

FIG. 46. DIVING FLAT INTO SHALLOW WATER

A dead silence would ensue till a few seconds later, when they would all come to the surface some twenty or thirty yards out, and start rolling about and splashing in the water, cleaning themselves and making sounds exactly like a lot of boys calling out and chaffing one another.

So extraordinary was this whole scene, that on first witnessing it we were overcome with astonishment, and it seemed to us almost impossible that the little creatures, whose antics we were watching, were actually birds and not human beings. Seemingly reluctant as they had been to enter the water, when once there they evinced every sign of enjoyment, and would stay in for hours at a time.

As may be imagined, the penguins spent a great deal of time on their way to and from the water, especially during the earlier period before the sea-ice had broken away from the ice-foot, as they had so far to walk before arriving at the open leads.

As a band of spotless bathers returning to the rookery, their white breasts and black backs glistening with a fine metallic lustre in the sunlight, met a dirty and bedraggled party on its way out from the nesting ground, frequently both would stop, and the clean and dirty mingle together and chatter with one another for some minutes. If they were not speaking words in some language of their own, their whole appearance belied them, and as they stood, some in pairs, some in groups of three or more, chattering amicably together, it became evident that they were sociable animals, glad to meet one another, and, like many men, pleased with the excuse to forget for a while their duties at home, where their mates were waiting to be relieved for their own spell off the nests.

After a variable period of this intercourse, the two parties would separate and continue on their respective ways, a clean stream issuing from the crowd in the direction of the rookery, a dirty one heading off towards the open water, but here it was seen that a few who had bathed and fed, and were already perhaps half-way home, had been persuaded to turn and accompany the others, and so back they would go again over the way they had come, to spend a few more hours in skylarking and splashing about in the sea.

FIG. 47. DIVING FLAT INTO SHALLOW WATER

In speaking of these games of the penguins, I wish to lay emphasis on the fact that these hours of relaxation play a large part in their lives during the advanced part of the breeding period. They would spend hours in playing at a sort of "touch last" on the sea-ice near the water's edge. They never played on the ground of the rookery itself, but only on the sea-ice and the ice-foot and in the water, and I may here mention another favourite pastime of theirs. I have said that the tide flowed past the rookery at the rate of some five or six knots. Small ice-floes are continually drifting past in the water, and as one of these arrived at the top of the ice-foot, it would be boarded by a crowd of penguins, sometimes until it could hold no more. (Fig. 39.) This "excursion boat," as we used to call it, would float its many occupants down the whole length of the ice-foot, and if it passed close to the edge, those that rode on the floes would shout at the knots of penguins gathered along the ice-foot (Fig. 40) who would shout at them in reply, so that a gay bantering seemed to accompany their passage past the rookery.

Arrived at the farther end, some half a mile lower down, those on the "excursion boat" had perforce to leave it, all plunging into the tide and swimming against this until they came to the top again, then boarded a fresh floe for another ride down. All day these floes, often crowded to their utmost capacity, would float past the rookery. Often a knot of hesitating penguins on the ice-foot, on being hailed by a babel of voices from a floe, would suddenly make the plunge, and all swim off to join their friends for the rest of the journey, and I have seen a floe so crowded that as a fresh party boarded it on one side, many were pushed off the other side into the water by the crush.

Once, as we stood watching the penguins bathing, one of them popped out of the water on to the ice with a large pebble in its mouth, which it had evidently fetched from the bottom. This surprised me, as the depth of the sea here was some ten fathoms at least. The bird simply dropped the stone on the ice and then dived in again, so that evidently he had gone to all the trouble of diving for the stone simply for the pleasure of doing it. Mr. J. H. Gurney, in his book on the gannet, says they (gannets) are said to have got themselves entangled in fishing-nets at a depth of 180 ft. and that their descent to a depth of 90 ft. is quite authentic, so that perhaps the depth of this penguin's dive was not an unusual one.

Fig. 48. Diving Flat

FIG. 49. ADÉLIES "PORPOISING"

The tide at the open water leads where they bathed ran a good six knots, but the Adélies swam quite easily against this without leaving the surface.

In the water, as on the land, they have two means of progression. The first is by swimming as a duck swims, excepting that they lie much lower in the water than a duck does, the top of the back being submerged, so that the neck sticks up out of the water. As their feet are very slightly webbed, they have not the advantages that a duck or gull has when swimming in this way, but supplement their foot-work by short quick strokes of their flippers. This they are easily able to do, owing to the depth to which the breast sinks in the water.

The second method is by "porpoising."

This consists in swimming under water, using the wings or "flippers" for propulsion, the action of these limbs being practically the same as they would be in flying. As their wings are beautifully shaped for swimming, and their pectoral muscles extraordinarily powerful, they attain great speed, besides which they are as nimble as fish, being able completely to double in their tracks in the flash of a moment. In porpoising, after travelling thirty feet or so under water, they rise from it, shooting clean out with an impetus that carries them a couple of yards in the air, then with an arch of the back they are head first into the water again, swimming a few more strokes, then out again, and so on.

I show a photograph of them doing this (Fig. 49).

Perhaps the most surprising feat of which the Adélie is capable is seen when it leaps from the water on to the ice. We saw this best later in the year when the sea-ice had broken away from the ice-foot, so that open water washed against the ice cliff bounding the land. This little cliff rose sheer from the water at first, but later, by the action of the waves, was under-cut for some six feet or more in places, so that the ledge of ice at the top hung forwards over the water. The height of most of this upper ledge varied from three to six feet.

FIG. 50. A PERFECT DIVE INTO DEEP WATER

Whilst in the water the penguins usually hunted and played in parties, just as they had entered it, though a fair number of solitary individuals were also to be seen. When a party had satisfied their appetites and their desire for play, they would swim to a distance of some thirty to forty yards from the ice-foot, when they might be seen all to stretch their necks up and take a good look at the proposed landing-place. Having done this, every bird would suddenly disappear beneath the surface, not a ripple showing which direction they had taken, till suddenly, sometimes in a bunch, sometimes in a stream, one after the other they would all shoot out of the water, clean up on to the top of the ice-foot. (Figs. 41 and 42.) Several times I measured the distance from the surface of the water to the ledge on which they landed, and the highest leap I recorded was exactly five feet. The "take off" was about four feet out from the edge, the whole of the necessary impetus being gained as the bird approached beneath the water.

The most important thing to note about this jumping from the water was the accuracy with which they invariably rose at precisely the right moment, the exact distance being judged during their momentary survey of a spot from a distance, before they dived beneath the water, and carried in their minds as they approached the ice. I am sure that this impression was all they had to guide them, as with a ripple on the water, and at the pace they were going, they could not possibly have seen their landing-place at all clearly as they approached it, besides which, in many cases, the ledge of ice on which they landed projected many feet forwards from the surface, yet I never saw them misjudge their distance so as to come up under the overhanging ledge.

During their approach they swam at an even distance of about three or four feet beneath the surface, projecting themselves upwards by a sudden upward bend of the body, at the same time using their tail as a helm, in the manner well shown in one of my photographs, in which one of the birds is seen in the air at the moment it left the water, the tail being bent sharply up towards the back.

Their quickness of perception is shown very well as they land on the ice. If the surface is composed of snow, and so affords them a good foothold, they throw their legs well forward and land on their feet, as shown in Figs. 41 and 42, but should they find themselves landing on a slippery ice-surface, they throw themselves forward, landing on their breasts in the tobogganing position as shown in Fig. 43.

The Adélies dive very beautifully. We did not see this at first, before the sea-ice had gone out, because to enter the water they had only to drop a few inches, but later, when entering from the ice terraces, we constantly saw them making the most graceful dives.

FIG. 51. SEA-LEOPARDS "LURK BENEATH THE OVERHANGING LEDGES OF THE ICE-FOOT, OUT OF SIGHT OF THE BIRDS OVERHEAD"

At the place where they most often went in, a long terrace of ice about six feet in height ran for some hundreds of yards along the edge of the water, and here, just as on the sea-ice, crowds would stand near the brink. When they had succeeded in pushing one of their number over, all would crane their necks over the edge (Fig. 44), and when they saw the pioneer safe in the water, the rest followed.

When diving into shallow water they fall flat (Figs. 45, 46, and 47), but into deep water, and from any considerable height, they assume the most perfect positions (Fig. 50) and make very little splash. Occasionally we saw them stand hesitating to dive at a height of some twenty feet, but generally they descended to some lower spot, and did not often dive from such a height, but twelve feet was no uncommon dive for them.

The reluctance shown by each individual of a party of intending bathers to be the first to enter the water may partly have been explained when, later on, we discovered that a large number of sea-leopards were gathered in the sea in the neighbourhood of the rookery to prey on the penguins. These formidable animals, of which I show some photographs, used to lurk beneath the overhanging ledges of the ice-foot, out of sight of the birds on the ice overhead. (Fig. 51.) They lay quite still in the water, only their heads protruding, until a party of Adélies would descend into the water almost on top of them, when with a sudden dash and snap of their great formidable jaws, they would secure one of the birds.

It seemed to me then, that all the chivvying and preliminaries which they went through before entering the water, arose mainly from a desire on the part of each penguin to get one of its neighbours to go in first in order to prove whether the coast was clear or not, though all this manœuvring was certainly taken very lightly, and quite in the nature of a game. This indeed was not surprising, for of all the animals of which I have had any experience, I think the Adélie penguin is the very bravest. The more we saw of them the fonder we became of them and the more we admired their indomitable courage. The appearance of a sea-leopard in their midst was the one thing that caused them any panic. With dozens of these enemies about they would gambol in the sea in the most light-hearted manner, but the appearance of one among them was the signal for a stampede, but even this was invariably gone through in an orderly manner with some show of reason, for, porpoising off in a clump, they at once spread themselves out, scattering in a fan-shaped formation as they sped away, instead of all following the same direction.

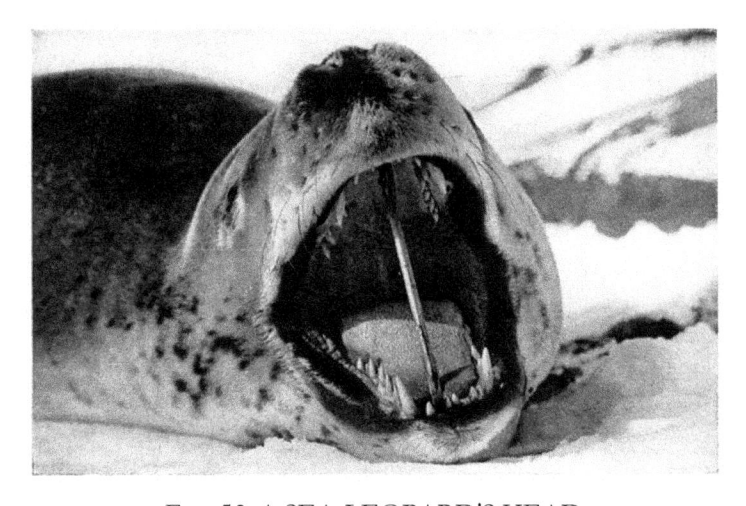

FIG. 52. A SEA-LEOPARD'S HEAD

As far as I could judge, however, the sea-leopards are a trifle faster in the water than the Adélies, as one of them occasionally would catch up with one of the fugitives, who then, realizing that speed alone would not avail him, started dodging from side to side, and sometimes swam rapidly round and round in a circle of about twelve feet diameter for a full minute or more, doubtless knowing that he was quicker in turning than his great heavy pursuer, but exhaustion would overtake him in the end, and we could see the head and jaws of the great sea-leopard rise to the surface as he grabbed his victim. The sight of a panic-stricken little Adélie tearing round and round in this manner was a sadly common sight late in the season.

Sea-leopards are no mean customers and should be treated with caution. Commander Campbell and I used to hunt them from a little Norwegian pram (a species of dinghy) which we rowed quietly up and down close under the ice-foot, shooting at the sea-leopards with a rifle when we saw their heads above water.

One day we had an interesting little adventure. We had shot and killed one, a fine bull about ten feet long, which had sunk to the bottom in some five fathoms. Having just pulled away from him, we were about ten yards from the ice-foot, when another very large sea-leopard overtook us, swimming from the direction of the dead bull. It passed under the pram, bumping against the keel in doing so. When about ten yards ahead of us it turned and made straight back for us, but as we were bows-on to it, it came right alongside the boat, churning up the water and wetting us. At this moment it turned on its side, its right fore-flipper beating the surface and its belly towards us, and was just starting to rear its head up when we both lunged at it with our paddles, and so pushed the little boat away from it. This brought us alongside the ice-foot, from which Campbell got a shot at it half a minute later, and wounded it in the neck. The moment after we lunged at it with our paddles it dived, then reappeared ten or fifteen yards off, rearing its head out of the water, and it was at this moment that Campbell shot it. After this it reappeared several times at the surface, but drifted away with the tide and we lost it.

Fig. 53. A Sea-Leopard 10 ft. 6½ ins. long

FIG. 54. A YOUNG SEA-LEOPARD ON SEA-ICE

The sea-leopard has not a reputation for attacking men in boats, and this one may have been actuated by curiosity merely, but in favour of its meaning to attack us were, first, that it came to us straight from the direction of the dead bull we had shot, and secondly, that it seems hardly likely that after bumping against our keel, mere curiosity could have tempted it to come back and try to look over the gunwale! As a rule we had to drift very quietly along when hunting sea-leopards, as the slightest sound frightened them away.

All that we could do to protect our friends was to shoot as many of these sea-leopards as possible but though we may have made some difference, there were always many about.

Some idea of the depredations committed by these animals may be gathered from the fact that in the stomach of one which we shot I found the bodies of eighteen penguins, in various stages of digestion, the beast's intestines being literally stuffed with the feathers remaining from the disintegration of many more. Photographs of these animals are seen in Figs. 52, 53, and 54.

Though the actual presence of a sea-leopard put the Adélies to confusion, causing them to "porpoise" madly away for a few hundred yards, yet once away from the immediate neighbourhood of the arch enemy, they appeared to think no more of him, and behaved as though there were no further need for anxiety, though probably they kept a sharp look-out nevertheless.

Evidence goes to show that the sea-leopard is the only living enemy, excepting man, that threatens the life of the adult Adélie penguin.

One day, as I watched some hundreds of Adélies bathing in an open lead, suddenly the back of an enormous killer-whale (*Orca gladiator*) rose above the surface as it crossed the lead from side to side, appearing from beneath the ice on one side and disappearing beneath it on the other. To my surprise, not the slightest fear was shown by the birds in the water. Had this beast been a sea-leopard, there would have been a stampede, and every bird have leapt from the water on to the sea-ice. On this evidence I formed the opinion that in all probability killer-whales do no harm to Adélie penguins; later I saw it confirmed, when a school of killers shaved close past several floes that were crowded with Adélies, and made not the least attempt to get at them, as they might so easily have done by upsetting the floes. Very probably this is because the agile bird can escape with such ease from the ponderous whale, and fears it no more than a terrier fears a cow, though he thinks twice before coming within reach of its jaws.

FIG. 55. "WITH GRACEFUL ARCHING OF HIS NECK, APPEARED TO ASSURE HER OF HIS READINESS TO TAKE CHARGE"

When the sea-ice had gone out, leaving open water right up to the ice-foot, a ledge of ice was left along the western side of the rookery, forming a sort of terrace or "front," with its sides composed of blue ice, rising sheer out of the water to a height of some six feet or more in places. From this point of vantage it was possible to stand and watch the penguins as they swam in the clear water below, and some idea was formed of their wonderful agility when swimming beneath the surface. As they propelled themselves along with powerful strokes of their wings, they swerved from side to side to secure the little prawn-like euphausia which literally swarm everywhere in the Antarctic seas, affording them ample food at all times. Their gluttonous habits here became very evident. They would gobble euphausia until they could hold no more, only to vomit the whole meal into the water as they swam, and so enlightened start to feast again. As they winged their way along, several feet beneath the surface, a milky cloud would suddenly issue from their mouths and drift slowly away down stream, as, without the slightest pause in their career, they dashed eagerly along in the hunt for more.

When a penguin returned to his mate on the nest, after his jaunt in the sea, much formality had to be gone through before he was allowed to take charge of the eggs. This ceremony of "relieving guard" almost invariably was observed.

Going up to his mate, with much graceful arching of his neck, he appeared to assure her in guttural tones of his readiness to take charge (Fig. 55). At this she would become very agitated, replying with raucous staccato notes, and refusing to budge from her position on the eggs. Then both would become angry for a while, arguing in a very heated manner, until at last she would rise, and, standing by the side of the nest, allow him to walk on to it, which he immediately did, and after carefully placing the eggs in position, sink down upon them, afterwards thrusting his bill beneath his breast to push them gently into a comfortable position. After staying by him for a little while, the other at length would go off to bathe and feed.

FIG. 56. "THE CHICKS BEGAN TO APPEAR"

(A TYPICAL GROUP OF NESTS)

FIG. 57. AN ADÉLIE BEING SICK

The length of time during which each bird was away varied considerably, but a "watch bill" was kept of one particular pair with the following result:(5)

Nov. 14. Egg laid. Hen sitting.

Nov. 27. A cock seen to join the hen for the first time since the 14th. He took her place on the nest. This was the first day on which any red guano was seen about the nest.

Dec. 10. The hen returned between 8 P.M. and 10 P.M., having been absent since November 27. Fresh red guano: the first for many days.

Dec. 14. The cock relieved the hen between 8 A.M. and 10 A.M.

Dec. 15. The hen relieved the cock between 8 A.M. and 10 A.M. Between 6 P.M. and 8 P.M. the chick was hatched, the hen remaining on the nest.

Dec. 17. At 8 A.M. the cock was found to have relieved the hen.

Dec. 18. Hen mounted guard between 6 P.M. and 8 P.M.

Dec. 20. Cock relieved guard about 8 A.M. At 8 P.M. both cock and hen were at the nest, the hen standing by it, the cock on it.

Dec. 21. The hen relieved guard at 8 P.M.

Dec. 23. Cock came back at noon and relieved guard.

Dec. 24. The cock remained on guard all day. The hen was gone from 1 P.M. till 6 P.M., when she returned and relieved guard.

Dec. 25. 8 A.M. Both at nest, hen still on.
10 A.M. changed guard. Hen gone.

Dec. 26. Hen on nest. Cock standing near.

Dec. 27. 8 A.M. Cock on nest.

Dec. 28. 8 A.M. Hen on nest.

Dec. 29. Cock relieved guard.

Dec. 30. Hen arrived 3 P.M. and relieved guard.

Dec. 31. 10 P.M. to midnight, changed, cock on. Both there at 10 P.M.

Jan. 1. 10 A.M. Both at nest.
12 noon. Both at nest. The youngster complicating matters by running away every time he was passed by the observer, thus getting himself and his parents embroiled with the neighbours.

Jan. 1. 2 P.M. Hen on nest. Cock gone.

Jan. 2. 10 A.M. Hen on nest.
12 noon. Chick disappeared.
2 P.M. Nest deserted.
4 P.M. Cock on nest. No chick.
8 P.M. Cock on nest. No chick.

Jan. 3. Cock on the nest with the chick.

FIG. 58. METHOD OF FEEDING THE YOUNG

From the above Table it will be seen that the hen was not relieved by the cock until a fortnight after she had laid her egg (in this case there was only one) so that probably she had been without food for a month. Then she left, and only returned to relieve the cock after the lapse of another fortnight, it being worth remarking that each was absent for the same length of time. When the chick was hatched, a different régime began, as of course the chick had to be fed and journeys to the sea made at regular intervals for the purpose of getting food.

When the chicks began to appear all over the rookery (Fig. 56), a marked change was noticed in the appearance of the parents as they made their way on foot from the water's edge to the nests. Hitherto they had been merely remarkable for their spotless and glistening plumage, but now they were bringing with them food for the young, and so distended were their stomachs with this, that they had to lean backward as they walked, to counterbalance their bulging bellies, and in consequence frequently tripped over the inequalities of the ground which were thus hidden from their gaze.

What with the exertion of tramping with their burden across the rookery, and perhaps on rare occasions one or two little disputes with other penguins by the way, frequently they were in some distress before they reached their destination, and quite commonly they would be sick and bring up the whole offering before they got there. Consequently, little red heaps of mashed up and half digested euphausia were to be seen about the rookery. Once I saw a penguin, after he had actually reached the nest, quite unable to wait for the chick to help itself in the usual manner, deposit the lot upon the ground in front of his mate. I saw what was coming and secured the accompanying photograph of the incident. (Fig. 57.) When this happens the food is wasted, as neither chick nor adult will touch it however hungry they may be, the former only feeding by the natural method of pushing his head down the throat of a parent, and so helping himself direct from the gullet. (Fig. 58.)

FIG. 59. PROFILE OF AN ADÉLIE CHICK

When the chicks are small they are kept completely covered by the parent who sits on the nest. They grow, however, at an enormous rate, gobbling vast quantities of food as it is brought to them, their elastic bellies seeming to have no limit to their capacity (Fig. 59); indeed, when standing, they rest on a sort of tripod, formed by the protuberant belly in front and the two feet behind.

I weighed a chick at intervals for some time, and this was the astonishing result:

	Ounces
The egg	4·56
The chick when hatched	3·00
Five days old	13·00
Six days old	15·75
Eight days old	24·75
Nine days old	28·50
Eleven days old	37·75
Twelve days old	42·50

To see an Adélie chick of a fortnight's growth trying to get itself covered by its mother is a most ludicrous sight. The most it can hope for is to get its head under cover, the rest of its body being exposed to the air; but the downy coat of the chick is close and warm, and suffices in all weathers to protect it from the cold. Fig. 60 illustrates what I have said very well, whilst Fig. 61 shows a mother with a chick twelve days old.

Whilst the chicks are small the two parents manage to keep them fed without much difficulty;(6) but as one of them has always to remain at the nest to keep the chicks warm, guard them from skuas and hooligan cocks, and prevent them from straying, only one is free to go for food. Later on, however, two other factors introduce themselves. The first of these is that the chick's downy coats become thick enough to protect them from cold without the warmth of the parent; and the second that as the chicks grow they require an ever-increasing quantity of food, and at the age of about a fortnight this demand becomes too great for one bird to cope with. At this time it is still necessary to prevent the chicks from straying and to protect them from the skuas and "hooligans," and so to meet these two demands a most interesting social system is developed. The individual care of the chicks by their parents is abandoned, and in place of this, colonies start to "pool" their offspring, which are herded together into clumps or "crèches," each of which is guarded by a few old birds, the rest being free to go and forage.

FIG. 60. A TASK BECOMING IMPOSSIBLE

It is quite likely that if a chick which has escaped from its own crèche joins another crèche it will get fed there, as it seems hardly possible for the adults to recognize the individuals of so large a gathering and to detect a stranger should one turn up, but there is good reason to believe that the old birds work for their own crèches only, and remain faithful to them for the rest of the season, because, as they make their way across the rookery, laden with the food they are bringing from the sea, it is sadly common to see them pursued by strayed and starving youngsters, plaintively piping their prayers for a meal; and these appeals are always made in vain, the old birds turning a deaf ear to the youngsters, who at last, weary and weak, give up the pursuit, and in the end fall a prey to the ever-watchful skuas. Further evidence is found in the fact that the chicks at the very back of the rookery and up at the top of the Cape are just as well nourished as those nearer the water, who are constantly passed by a stream of food-laden parents.

Twice already I have mentioned that strayed chicks fall a prey to "hooligan" cocks. These hang about the rookery often in little bands. At the beginning of the season there are very few of them, but later they increase greatly, do much damage, and cause a great deal of annoyance to the peaceful inhabitants. The few to be found at first probably are cocks who have not succeeded in finding mates, and consequently are "at a loose end." Later on, as their numbers are so greatly increased, they must be widowers, whose mates have lost their lives in one way or another.

Many of the colonies, especially those nearer the water, are plagued by little knots of "hooligans," who hang about their outskirts, and should a chick go astray it stands a good chance of losing its life at their hands. The crimes which they commit are such as to find no place in this book, but it is interesting indeed to note that, when nature intends them to find employment, these birds, like men degenerate in idleness.

Some way back I made some allusion to the way in which many of the penguins were choosing sites up the precipitous sides of the Cape at the back of the rookery. Later I came to the conclusion that this was purely the result of their love of climbing. There was one colony at the very summit of the Cape,[7] whose inhabitants could only reach their nests by a long and trying climb to the top and then a walk of some hundred yards across a steep snow slope hanging over the very brink of a sheer drop of seven hundred feet on to the sea-ice.

FIG. 61. ADÉLIE WITH CHICK TWELVE DAYS OLD

During the whole of the time when they were rearing their young, these mountaineers had to make several journeys during each twenty-four hours to carry their enormous bellyfuls of euphausia all the way from the sea to their young on the nests—a weary climb for their little legs and bulky bodies. The greater number who had undertaken this did so at a time when there were ample spaces unoccupied in the most eligible parts of the rookery.

I have mentioned that large masses of ice were stranded by the sea along the shores of the rookery. These fragments of bergs, some of them fifteen to twenty feet in height, formed a miniature mountain range along the shore. All day parties of penguins were to be seen assiduously climbing the steep sides of this little range. Time after time, when half way up, they would descend to try another route, and often when with much pains one had scaled a slippery incline, he would come sliding to the bottom, only to pick himself up and have another try. (Fig. 63.)

Generally, this climbing was done by small parties who had clubbed together, as they generally do, from social inclination. It was not unusual for a little band of climbers to take as much as an hour or more over climbing to the summit. Arrived at the top they would spend a variable period there, sometimes descending at once, sometimes spending a considerable time there, gazing contentedly about them, or peering over the edge to chatter with other parties below.

Again, about half a mile from the beach, a large berg some one hundred feet in height was grounded in fairly deep water, accessible at first over the sea-ice, but later, when this had gone, surrounded by open water. Its sides were sheer except on one side, which sloped steeply from the water's edge to the top.

From the time when they first went to the sea to feed until the end of the season, there was a continual stream of penguins ascending and descending the berg. As I watched them through glasses I saw that they had worn deep paths in the snow from base to summit. They had absolutely nothing to gain by going to all this trouble but the pleasure they seemed to derive from the climb, and when at the top, merely had a good look round and came down again.

FIG. 62. A COUPLE WITH THEIR CHICKS

When the birds were arriving at the rookery I watched for those who were to nest up the cliff, and several times saw birds on arriving at the rookery make for the heights without any hesitation, threading their way almost in a straight line through the nests to the screes at the bottom of the cliff, and up these to one or other of the paths leading up its side. Probably these had been hatched there, or had nested there before, and were making their way to their old haunts, but my notes on their nesting habits go to show that the cocks, at any rate, cannot keep to the same spot during successive years. It is the hen who chooses the site, and stays on it, as I have shown, until a mate comes to her, and wins her, very often only after defeating many other competitors.

The waste of life in an Adélie rookery is very great, and is due to the following causes:

- *The eggs.*
  - Skuas.
  - Cocks fighting among the nests.
  - Floods from thaw water.
  - Death of parents.
  - Snow-drifts.
  - Landslides.
- *The young chicks.*
  - Skuas.
  - Landslides.
  - "Hooligan" cocks.
  - Getting lost.
  - Death of parents.
- *Adults.*
  - Sea-leopards.
  - Landslides.
  - Snow-drifts.

In the above lists I have made no mention of the wanton depredations committed—owing to the licence given to ignorant seamen—by expeditions which visit the Antarctic from time to time, but as these visits are made at rare intervals, they cannot greatly affect the population.

FIG. 63. ADÉLIE PENGUINS HAVE A STRONG LOVE OF CLIMBING FOR ITS OWN SAKE

Some of the items in my list require explanation. The screes at the foot of the cliff at Cape Adare are perhaps the most thickly populated part of the rookery. As the thaw proceeds, boulders of different sizes are continually falling down the cliff, some of them for many hundreds of feet before they finally plunge in among the nests on the screes, doing terrible damage, and often rolling some distance out into the rookery. At other times, owing to the bursting out of thaw water which has been dammed up at the top of the cliff, large landslides are caused which bury many hundreds of nests beneath them. In fact, these screes on which the nests are built have been formed by these landslides taking place from year to year, and no doubt form the graves of thousands upon thousands of former generations. One of these slides took place whilst we were at the rookery, doing terrible damage. A crowded colony of Adélies were nesting just below, and the avalanche passed right through and over them, causing the most sad havoc. We found hundreds of injured and dying, some of them in a pitiable condition. Several were completely disembowelled, others had the whole skin of their backs torn down and hanging behind them in a flap, exposing the bare flesh. Dozens had broken or dislocated legs and flippers.

The worst feature was that many were buried alive beneath the snow, or pinned down to the ground by masses of basalt. Twice I saw one flipper sticking out of the snow, moving dismally, and dug out in each case a badly injured bird which would have lingered perhaps for days, because loose snow does not always suffocate, owing to the amount of air contained in its interstices, and to the fact that diffusion takes place through it very readily.

Several of us spent a long time in killing with ice-axes those that seemed too badly injured to recover.

It was remarkable to see the way in which all the nests which had escaped the avalanche, however narrowly, were still sat upon by their occupants, as if nothing had happened. Also I saw several badly injured birds sitting on their eggs, some of them soaked in blood, so that they looked like crimson parrots. The amount of bloodshed must have been great, as the snow was dyed with blood in all directions. As a cascade of water followed the avalanche, and continued for some hours, spreading out into little rivers among the nests, many were being deluged, and some of the penguins actually were sitting in the running water, in a vain attempt to keep warm their drowned chicks and spoiled eggs.

Sometimes, digging at hazard in the drifted snow, I came on birds that had been deeply buried, and though they were held down and kept motionless by the weight of the snow covering them, most of them were alive, and I have no doubt many dozens died a lingering death in this way. Such as had merely suffered broken flippers or legs, I spared, and the next day nearly all of these seemed to be doing well. One bird I found sitting on two eggs which were in the middle of a rivulet of water, so I lifted them out and put them on dry ground close by, but the parent would have nothing to do with them after this.

Fig. 64. ADÉLIES ON THE ICE-FOOT

A feature of the above scene, which one could not help noticing, was that however badly a penguin was injured it was never molested by the others, as is almost invariably the rule among other birds, including their near neighbours the skuas. I have seen a sick skua hunted continuously for over an hour by a mob of its own kind who would not allow it to settle on the ice for a moment's rest.

Another item of my list requiring explanation is "snow-drifts."

During both spring and summer there are occasional snowstorms, and during these the birds sit tight on their nests, sometimes being covered up by drift. As a rule the bird on the nest keeps a space open by poking its head upwards through the snow, but sometimes it becomes completely buried. Air diffuses so rapidly through snow that death does not take place by suffocation, and the bird can live for weeks beneath a drift, sitting on its nest in the little chamber which it has thawed out by its own warmth. Generally after a few hours the snow abates and settles down sufficiently to expose the nest once more, but sometimes a breeze springs up which is not strong enough to blow the snow away, but simply hardens the surface of the drift into a crust which lasts for several weeks, and the birds are imprisoned in consequence. Then little black dots are seen about the surface of the drift, which are the heads of penguins thrust through their breathing holes.

On one such occasion I witnessed an interesting little incident. An imprisoned hen was poking her neck up through her breathing hole when her mate spied her and came up. He appeared to be very angry with her for remaining so long on the nest, being unable to grasp the reason, and after swearing at her for some time he started to peck at her head, she retaliating as far as her cramped position would allow. When she withdrew her head, he thrust his down the hole till she drove it out again, and as this state of things seemed to be going on indefinitely, I came up and loosened the crust of snow which imprisoned her, on which she burst out, and seemed glad to do so. She was covered with mire, having for many days been sitting in a pool of thaw water which had swamped her nest and evidently spoilt the eggs. When I put her back on the nest, she sat there for some time, but eventually they both deserted. I should say that some hundreds of nests were spoilt in this way.

Fig. 65. ADÉLIES ON THE ICE-FOOT

As I photographed the above incident at intervals, different stages are seen on Figs. 67 to 69.

I have mentioned that eggs got lost owing to cocks fighting among the nests. When hens are incubating the eggs they never leave the nest under any circumstances until relieved by their mates, being most reliable and faithful to their charge. They squabble continually with their nearest neighbours, whom they seem to hate, but only retaliate on those within reach, using their bills only to peck at each other's heads without shifting their position.

The cocks, however, are less dependable. Starting a quarrel in the same way as the hens do, their militant instincts soon became aroused, on which they are apt to jump up and start a furious fight with flippers, staggering to and fro over their nests, and very often spilling the eggs, which are lost in consequence. On certain occasions I was able to interfere between the combatants, and replace the eggs, on which they would return to their domestic duties and seem to forget the incident. A good many eggs must have been lost in this way during the season.

Late in the season an occurrence took place for which I have never been able to find any explanation. Occasionally I had noticed that the penguins had crowded together more than usual on the ice-foot, multitudes of them standing for hours without any apparent purpose. A good idea of this scene may be got from the frontispiece.

One morning Mr. Priestley came into the hut and told me that "the penguins were drilling on the sea-ice," and that I had better come and look at them. I went with him to the ice-foot, and this is what we saw.

Many thousands of Adélies were on the sea-ice between the ice-foot and the open water leads, then some quarter of a mile distant. Near the ice-foot they were congregating in little bands of a few dozen, whilst farther out near the water, massed bands some thousands strong stood silent and motionless. Both the small and the large bands kept an almost rectangular formation, whilst in each band all the birds faced the same way, though different bands faced in different directions.

4

Fig. 66. ADÉLIES ON THE ICE-FOOT

As we watched it became evident that something very unusual was going on. First, from one of the small bands, a single bird suddenly appeared, ran a few yards in the direction of another small band, and stopped. In the flash of a moment the entire band from which he came executed the movement "left turn," this bringing them all into a position facing him. So well ordered was the movement that we could scarcely believe our eyes. Then from the small band our single bird had approached, another single bird ran out, upon which his own party did exactly as the first had done, so that the two bands now stood facing one another, some fifteen yards apart.

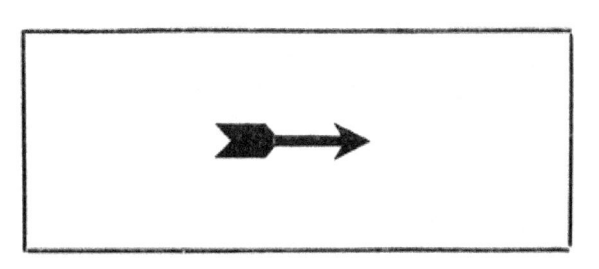

ARROW SHOWS DIRECTION IN WHICH ALL THE BIRDS FACED

Then spontaneously, the two bands marched straight toward one another, and joined to form one body. After this we saw the same procedure being enacted in many other places, the penguins coming down from the rookery and forming small bands which joined together. Then the augmented bodies would join other augmented bodies, to form still larger ones, which then joined together, and so on until a great mass of birds stood together in rows all facing in one direction like a regiment of soldiers. One of these masses stood not far from us, a compact rectangular gathering, as shown on page 109.

They stood thus for a long time, quite motionless and silent, when suddenly as before, a single bird darted out from among the crowd and ran a few yards toward the open water, when, as if it had received a word of command, every bird faced left as in the diagram below.

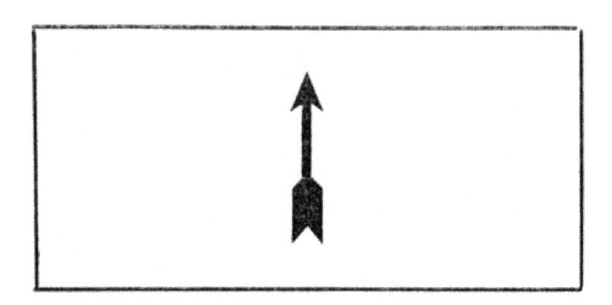

ARROW SHOWS DIRECTION IN WHICH ALL THE BIRDS FACED

After this the whole crowd marched for the water, keeping its formation almost unchanged till it arrived at the edge of the ice, when it halted, and subsequently entered the water in batches.

Fig. 67. "An Imprisoned Hen was Poking her Head up through her Breathing Hole"

Fig. 68. "Her Mate Appeared to be Very Angry with Her, Being Unable to Grasp the Reason Why She Could Not Come Off the Nest"

This procedure continued for many hours, the penguins that day observing this extraordinary behaviour, the most astonishing part of which lay in the accuracy of their drill-like movements, so that we might have been watching a lot of soldiers on parade. Perhaps the sudden motions of these bodies of birds were brought about by a sound uttered by the single bird which acted as leader, though we did not hear this. The actual reason for this departure from their usual customs is beyond my knowledge. There was nothing to be seen to account for it, but the penguins evidently obeyed some instinct which affected them all on this and two subsequent occasions, when the same thing took place.

My own idea is that in former times the penguins used to mass together as other birds do, before their annual migration, perhaps as far back as the day when their wings were adapted for flight, and that the phenomenon described above was a relic of their bygone instincts.

When the chicks' down has been moulted and their plumage acquired, they proceed to the water's edge and here they learn to swim.

In the autumn of 1912, at a small rookery which I came upon on Inexpressible Island, I had an opportunity of watching their first attempts in this direction. Crowds of young Adélies were to be seen on the pebbly beach below their rookery, much of the ice having disappeared at this late season, leaving bare patches of shingle which were very suitable for the first swimming lesson.

Many old birds paddled in for a short distance, and crouching in a few inches of water, splashed about with their flippers to give the youngsters a lead. Some of the latter needed little encouragement, and took readily to the strange element, very soon swimming about in deep water, but others seemed more timid, and these latter were urged in every possible way by the old birds, some of whom could be seen walking in and out of the water, and so doing what they could to give their charges confidence.

In this duty one or two old birds might be seen with a little crowd of youngsters, so that evidently the social instincts which gave rise to the crèche system in the first place were extended to the tuition of the young and thus to their preparation for the journey north.

FIG. 69. "When she broke out, they became reconciled"

FIG. 70. Adélie's Nests on Top of Cape Adare, to reach which they must make a Precipitous Climb of 1000 feet

Up in the rookery, fully fledged youngsters could be seen clamouring in vain for food, the old birds resolutely refusing to feed them now that they were able to forage for themselves. The adults who instructed the young in the water had finished their moult, and were themselves ready to depart. Many others, however, still wandered disconsolately about the land, some of them only half fledged, and moping under boulders or any sort of shelter from the chilly breezes, and long after all the youngsters had departed, solitary moulting birds were to be found, emaciated and miserable, patches of loose feathers still clinging to the new coat which was making such a tardy growth. In some places we found these old birds in holes under the rocks, the old moulted feathers making some sort of a bed which helped to protect their late wearers from the cold.

Both at Cape Adare in 1910 and at Inexpressible Island in 1911, I found that though young and old left the rookery simultaneously at first, yet after all the young had departed many adults still remained behind owing to the lateness of their moult, and this is directly at variance with the remarks of Mr. Borchgravink on the subject, because he says that the old birds all leave the rookery first, abandoning the young, who are driven by necessity to take to the water and learn to swim.

Well indeed was it for my companions and me that this was so, for in the autumn of 1912 we were in sore straits for food, and had it not been that at a very late date we collected some ninety old moulting birds on Inexpressible Island, I doubt if we would have seen the sun rise in the next spring.

At Cape Adare in 1911, half the rookery had departed when we arrived in the autumn. The rest took to the sea in batches some hundreds strong. These parties wandered about the beach and ice-foot in company for some time, then entering the water and swimming northward they were seen no more.

Those that moulted sometimes remained solitary whilst in the acuter stages, but nevertheless moulting parties often were seen looking very miserable, doubtless feeling in their unprotected state the effects of winds which were getting keener and more severe now that the sun was departing.

When all the youngsters had gone, some thousands of old birds still remained, and waited for many days after they had acquired their full plumage before they left. Then these in time disappeared, leaving the rookery empty and desolate. On March 12 I photographed the last party: all black-throated adults. Two days later a couple appeared on the beach, apparently having come back for a last look at us. Then these, too, disappeared, and as we looked at the empty silent beach we could not help contrasting it with the noise and bustle of a short time ago.

The last penguin had gone, and the sun disappearing below the horizon, left us alone with the Antarctic night.

APPENDIX

APPENDIX

(A) Plumage and Soft Parts.

The following description of the plumage and soft parts of *Pygoscelis Adeliæ*, which is perfectly correct, is taken from the zoological report of the *Discovery* Expedition.

Soft Parts.

"*Bill*, when first hatched, blackish. A week old, black terminally, deep red at the gape and along the cutting edges. Immature of the first year, blackish. Adult, brick-red, the upper bill black terminally, and the mandible black along the cutting edge.

"*Iris*, brown; varying between reddish brown and greenish brown.

"*Eyelids*, black throughout the first year; pure white in the adult at fourteen months and onwards.

"*Feet*, flesh red; dusky when first hatched, brightening in the first week or two. Immature and adult, pale flesh pink above, black beneath (in some cases piebald beneath).

"*Claws*, brown."

In the majority of the chicks the down is uniformly dark and sooty, but here and there, in progeny of quite normal parents, one may find nestlings of so pale a grey as to be almost silvery white, with blackish heads, possibly a reversion to an earlier type, and, at any rate, suggestive of the young of the Emperor penguin, which perhaps represents the oldest stock of all. According to Dr. Bowdler Sharp, the colour of the head is in all cases blacker in the earlier stages than the rest of the body.(8)

As the chick ages the colour of its down changes, and all of it takes on a dull rusty brown colour. As it moults the abdomen and thighs change first, and white feathers appear in place of the down. Then come changes on the head, round the bill, and at the tail; the upper breast, neck, and back being the last parts to moult.

The feet, which in the young nestling have been almost black, change in colour to a brick-red that shows up very markedly against the rusty brown down, looking as if the legs were raw and inflamed. Later the permanent flesh colour is acquired, with black plantar surfaces. The nails are black at first, and later change to brown.

When the nestling down is shed, the resulting plumage is that of the adult, except that the throat is white instead of black. The upper part of the head and neck are bluish black, the throat, fore-neck, breast, and abdomen being a pure dazzling metallic white, a sharp line separating the white from the black areas. The flippers are the same bluish "tar" black on the back and white beneath.

In addition to the distinctive pure white plumage of the throat, the immature bird differs from the adult in one very marked particular, which is that the eyelids are black, as in the chick, and do not acquire the staring whiteness which is so distinctive of the adult Adélie penguin, increasing, as it does, the white area of the sclerotic so that the bird has the appearance of being perpetually surprised or very angry.

The iris is a rich reddish brown in the adults, but variable in the young.

At Cape Adare the light grey "silvery" coloured chicks mentioned by Dr. Wilson were by no means uncommon; in fact, quite a large proportion of the chicks had very light-coloured down. This is shown in some of the specimens I brought back to the British Museum.

(B) VARIATIONS.

Variations occasionally are met with in the plumage and soft parts of Adélies. The least rare of these consist of tufts of white feathers amongst the black plumage of the head. Several specimens so marked were seen at Cape Adare during the summer of 1911–12.

When these white tufts were present the feathers comprising them were usually longer than the black feathers among which they appeared, so that they stuck out in an untidy manner, and were very conspicuous.

In marked distinction to the slight variations above described were the three "Isabelline" varieties that I preserved, and are now to be seen in the British Museum collection. As these variations are very startling, and of the greatest interest, I give below a full description of their plumage and soft parts.

First specimen captured on the Cape Adare rookery on November 4, 1911.

Iris, light brown. *Eyelids*, white. *Bill*, light brown. *Feet*, white. *Claws*, light brown.

The whole of the area covered by black feathers in the normal bird was covered by those of a very light fawn, somewhat darker on the neck and shoulders than elsewhere. *Sex*, male.

Second specimen captured on November 14, 1911.

Iris, light brown. *Eyelids*, white. *Bill*, light brown; mandible, blackish on dorsum; maxilla, blackish on cutting edges. *Feet*, white on both surfaces. *Claws*, light brown.

In place of the black feathers of the normal bird, there was a fawn-coloured plumage, darkest on head and neck; lightest at bottom of back, back of flippers, and on shoulders.

Sex, female.

Third specimen captured on December 23, 1911.

Iris, light brown. *Feet*, browny white. *Claws*, brown. *Bill*, brown; very dark on dorsum of mandible. *Eyelids*, white with a pink tinge.

In place of the black feathers of the normal bird, this specimen had those of a very light cream colour: in fact very slightly darker than the white area but deepening in shade to light fawn on the head, neck, and shoulders.

Sex, male.

The second specimen had mated with a normal cock. In each case the Isabelline birds were very much more docile than the normal forms. For instance, they did not struggle when picked up, as the others would have done, and the third specimen, when brought into our hut, gazed around with curiosity and apparent contentment, and showed not the least resentment at its captivity. A normal bird would have struggled and fought to the last extremity. Each bird was killed with chloroform.

So carefully did we keep the entire rookery under observation that I do not think it likely there were any more Isabelline forms. Thus we can conclude 3/750000 roughly represents the proportion of Isabelline forms among the species.

PART III

McCORMICK'S SKUA GULL

A BOOK which treats of Adélie penguins scarcely can be complete without reference to the beautiful McCormick's skua gull (*Megalestris maccormicki*), as probably no Adélie rookery exists without its attendant band of skuas, who build their own nests very close to and occasionally among those of the penguins on whom they prey, almost entirely supporting themselves and their young upon the eggs and young offspring of their hosts.

Mention has been made of these birds from time to time through the previous pages, and some idea of their habits already will have been formed. In point of fearlessness they fall somewhat short of the Adélie, but exhibit, perhaps, rather more caution in their dealings with man than the gulls who visit St. James's Park in London and are fed by the children there, frequently from the hand, though probably in a very few days they might become extremely tame were their short experience of mankind made less bitter. The majority of explorers, like most men, though kindly by nature, are entirely thoughtless in their dealings with wild animals, and the skuas approach them only to be killed or severely injured by the ice-axes or rocks that are thrown at them in wanton sport as they light on the ground or hover near the visitors, whom they quickly discover to be their bitter and relentless foes.

Arriving at the rookeries somewhat later than the Adélies, they do not lay their eggs until the beginning of December. Practically no nest is made, a mere hollow being worked in the ground, in which the bird sits. Frequently several hollows are made before the hen finally settles where she will lay. The two eggs, which are brownish olive thickly and darkly mottled with brown, are incubated for four weeks, after which the chicks are hatched.

From the moment of their first appearance from the egg these chicks exhibit the most extraordinary precocity. Covered with pale slaty-grey down, they look anything but the pugnacious little animals they turn out to be. Their one idea, besides feeding, seems to be to fight one another, and they may be seen to roll about the nest, locked together, fighting with beak and claw. They are fed from the ground, and may be seen picking about among the stones like the little domestic chickens, which they very much resemble. After a time invariably one of the chicks disappears, and as dead youngsters are not to be found, they are probably eaten by neighbours who have caught them wandering; in fact, Mr. Ferrar, of Captain Scott's first expedition, actually saw a Skua pick up a wandering chick of its own species and fly off with it, followed by a screaming flock of its neighbours, who sought to rob it of its prey.

In order to find out how many eggs a Skua would lay, I marked some nests, and took the eggs as they were laid. In each case a second egg was laid, but when this was taken no more appeared. In two nests I removed the first egg as soon as it was laid, but left the second, which was then sat upon by the parent, who was content with it, or unable to lay a third.

When any of us approached their nests the old birds would fly round in wide circles, making wild "stoops" at our heads each time they passed over us, in the evident attempt to frighten us away. Occasionally they would actually knock our heads with a wing, and nothing seeming to scare them off, they would swoop past us time after time in a most disconcerting manner. In order to keep them at a distance without having to keep a constant look-out, when I was in the neighbourhood of their nest, I used to walk about holding a ski-stick or the handle of my ice-axe straight above me, and they would swoop at the top of this instead of my head, which was infinitely preferable. One day when a high wind was blowing on top of Cape Adare, I had my ice-axe knocked clean out of my hand by one of the Skuas flying straight into the handle, the heavy blow seeming to affect the bird but slightly.

There was a "skuary" on the screes, close to a thickly populated part of the rookery, but the majority of these birds made their nests right at the top of Cape Adare, from which point of vantage they surveyed the entire rookery, and a very sharp look-out they kept too, for no sooner did we start to flense a seal than a flock of them descended to gobble at the lumps of blubber as we threw them on the ground. In this occupation they exhibited the greatest jealousy, and when there was a hundred times as much blubber on the ground as all the skuas possibly could have eaten, they continually tried to drive each other away. When fighting they rarely stayed on the ground, but leapt at one another into the air, and one of the illustrations shows two Skuas in the act of doing this. Their great spread of wing is well shown in this photograph. (Fig. 71.)

When penguins' eggs were plentiful in the rookery the Skuas flew very low over the ground, and as they passed over each colony of nests the sitting birds would crouch low upon them, a very necessary precaution, as I have described already in these pages the unerring way in which the Skuas picked up the penguins' eggs when they were left uncovered. Broken and empty shells strewed the ground in the vicinity of the Skuas' nests, and it is probable that in a large rookery, such as that at Cape Adare, thousands of eggs are destroyed by them annually.

The instinct of the thief is most strongly marked in the Skua tribe, and I am afraid that the mere love of thieving alone actuates them on many occasions. For instance, when I was skinning a seal one day near Cape Evans I left a pair of field-glasses lying on a coat close by, and on looking round saw a Skua in the act of making off with them, holding them by the strap in his beak. A sudden yell caused the offender to drop the glasses, fortunately when they were but a yard from the ground. Again, when the crew of an Antarctic ship were engaged in blasting the sea-ice which imprisoned it, a Skua flew off with one of the detonators which had been left on the ice. I think the detonator contained dynamite, but at any rate I am told that there was a stampede on the part of the men to get away from under the bird as it flew overhead!

When, with two companions, I visited a skuary at the back of the penguin rookery at Cape Royds, the Skuas circled over us in a way I have described above, but instead of swooping at our heads, some of them repeatedly dropped their guano on to us as they passed over, timing the process with such surprising accuracy that I was hit once, and Commander Campbell no less than three times. The following year when at Cape Adare, I expected the same treatment from the Skuas there, but curiously enough, these never did it. That one skuary should have adopted such tactics and another not, is a very curious thing, but it may possibly be that the Cape Royds Skuas discovered the trick during the stay of Sir Ernest Shackleton's expedition, who had spent a year there quite recently, and of the *Discovery* expedition which spent two years at Hut Point, but a few miles distant, whereas the only men who ever inhabited Cape Adare wintered there some fifteen years before. But this is mere speculation.

FIG. 71. "LEAPT AT ONE ANOTHER INTO THE AIR"

FIG. 72. A SKUA BY ITS CHICK

When one of the parent Skuas is on the ground near its nest, on the approach of anyone it throws its head back, opens its wings, and loudly proclaims its whereabouts with its raucous cawing notes. When hovering over food, and at other times when not alarmed or angry, the sounds made by a Skua are very like those of the common Herring Gull, and not altogether unmusical at times, especially when making the little shrill piping note, by which I have often thought that gulls so nearly imitate the squeaking of a block in its sheaf.

When the penguin chicks are hatched, the Skuas prey upon these in a most cruel manner, and should a chick wander away from the protecting old birds, a Skua is almost certain to pounce upon and kill it. This it does by pecking its eyes out, after which, with powerful strokes of its beak, it gets to work on its back and quickly devours the kidneys.

The dead bodies of hundreds of chicks are seen strewn about the rookery, and especially in the neighbourhood of the Skuas' nests, as very often they carry them there. All these dead chicks are seen to have two holes picked through their backs, one on each side, corresponding to the position of each kidney.

Besides the penguins' eggs and young, there is another fruitful source of food for the Skuas to be found along the Antarctic coasts at the early part of the year, and that is during the time when the seals are bringing forth their young upon the sea-ice. The Skuas attend upon them then, and devour the afterbirths. In the second volume of the *Discovery* reports Dr. Wilson mentions that large numbers of Skuas were noticed at Granite Harbour, and I have no doubt that they had congregated there for this purpose, as when passing the spot on a spring journey along the sea-ice in 1912, we saw many hundreds of Weddell seals with their young. So many were there, that as we lay in our sleeping-bags during the night, the bleating of the little calves near our tents conveyed to our half-awakened senses the impression that we were in the midst of lambing fields at home!

The soft parts are coloured as follows:

Bill, black.

Iris, dark brown.

Legs, *toes*, and *webbs*, black, excepting a patch of bright blue just above the tibio-metatarsal joint, in young fledglings. Wholly black afterwards. (They have a very fine spread of webb.)

Claws, black.

The feathers of the head, neck, and breast vary from very light buff, or almost white, to rich dark brown.

	SKUAS' TIME TABLE		
	McMurdo Sound		Cape Adare
	1902	1903	1911
1st bird arrived	Nov. 3	Oct. 25	Oct. 26
1st egg seen	Dec. 9	Dec. 2	Nov. 29
1st chick hatched		Jan. 1	
Last bird left	Mar. 30	Apr. 7	

A SHORT NOTE ON EMPEROR PENGUINS(9)

THE Emperor is by far the largest of all penguins, weighing between 80 and 90 lbs. It is also a particularly handsome and graceful bird. By nature it seems much like the Adélie, except that its general demeanour is extremely dignified, and its gait, as it approaches you over the snow, slow and deliberate.

The most marked difference in the habits of the Adélie and the Emperor lies in the respective seasons at which each lays and incubates its eggs. Unlike the Adélie, which, as we have seen, chooses the warmest and lightest months of the year for the rearing of its young, the Emperor performs this duty in the darkest, coldest, and most tempestuous time. The only reason that has been suggested for this custom is that many months must pass before the chicks are fully fledged. Were they hatched in December (midsummer) as are Adélies, autumn would find them still unfledged, and probably they would perish in consequence, whereas, being hatched in the early spring, they are fostered by their parents until the warmer weather begins, and then have the entire summer in which to accomplish their change of plumage.

FIG. 73. AN EMPEROR PENGUIN

The only Emperor rookery known to man at the present day was discovered by Lieuts. Royds and Skelton, of Captain Scott's first Antarctic expedition, on the sea-ice beneath Cape Crozier. Here in the dark days of July this extraordinary bird lays its one egg upon the ice.

In the winter of 1911 a very brave journey was made to this spot by a party of Captain Scott's officers, consisting of Dr. Wilson, Lieut. Bowers and Mr. Cherry-Garrard. The experiences of this little band were so terrible that it is remarkable they ever returned to tell of them. Temperatures of −78° F. were encountered, and the most severe blizzards at lower temperatures than any sledging-party had yet endured. Under these truly terrible conditions the Emperors lay their eggs and hatch their young.

The mortality under such circumstances is very high, as one would expect. Avalanches of ice fall from the cliffs above, crushing many of the parent birds, and causing hundreds of eggs to be deserted. As Dr. Wilson stated, the ice cliffs beneath which these remarkable animals sat were so unstable that no man in his senses would camp for a single night beneath them. In spite of this, evidence showed that after an avalanche of ice blocks from above, which had caused some of the Emperors to leave their eggs on the ice and bolt in terror, many of them had returned and continued to sit on the eggs which had been frozen and killed by the frost in their absence, continuing to do so long after they were completely rotten. Indeed, in their desire for something to hatch, some who had been deprived of their eggs, were seen to be attempting to incubate pieces of ice in their place, and, unlike Adélies, they seem ever ready to snatch and foster the young of their neighbours.

The first time the rookery at Cape Crozier was visited, not above one thousand birds occupied it. On the second occasion their numbers were far short of this. By the springtime only one out of ten or twelve birds are seen to be rearing young, so it is obvious other rookeries await discovery in other parts, as there are a large number of Emperors to be seen along the Antarctic coasts.

FIG. 74. PROFILE OF AN EMPEROR

When in the *Terra Nova* we made our way along the face of the great Barrier to the eastward, we saw large numbers of Emperors, especially to the extreme eastward where a heavy hang of pack-ice blocked our further passage, and I have little doubt that future exploration will disclose a rookery or rookeries in this direction.

Again, in the spring of 1912, when nearing the end of a sledge journey from the northward to Cape Evans we passed large gatherings of Emperor penguins on some very old sea-ice under the Barrier's edge, along the southern end of McMurdo Sound, and it seems not at all unlikely that they may breed here too. Unfortunately we were unable at that time to make detours, so had to leave the question unsettled, but if they do breed here, they must have far to go to get food during those winters when the sea-ice does not break out of the Sound.

The growth of the Emperor chick is slow, when compared with the mushroom-like rate at which the Adélie youngster increases its substance.

Approximately the egg is laid at the beginning of July and hatched out some seven or eight weeks later. During the period of incubation, which duty is shared by all, male and female alike, the egg is held in a loose fold of skin at the lower part of the abdomen, the skin of the adults being worn bare of feathers in this region.

When hatched out, the chick is coveted by every unoccupied adult, and so desperate at times are the struggles for its possession that very frequently it gets injured or killed by its would-be foster parents. Dr. Wilson has estimated the mortality among the chicks before they shed their down at 77 per cent. and thinks that half this number are killed by kindness. Very often, in fact, they will crawl under projecting ledges of ice, or anywhere to escape the attentions of half a dozen or so of adults, all bearing down upon them together, only to meet and struggle for their possession, during which process the innocent cause may get trampled and clawed to death. So strong is the maternal instinct of the Emperor, that frozen and lifeless chicks are carried about and nursed until their down is worn away. In fact, the scientists who visited the rookery were unable to get good specimens of dead chicks, as all of these had been treated in this way.

Fortunately the Emperor chick escapes the depredation of the Skua gull, which plays such havoc in the Adélie rookeries, because the Skua does not come south until the summer, by which time the Emperor chicks are well grown. As in the case of the Adélies, the black throat is not acquired until the second moult. When this has taken place, the bird looks remarkably handsome. The bill, which is curved and tapering, is bluish black, but the posterior half of the mandible is coloured a beautiful lilac. The head and throat are black, whilst on each side of the neck is a patch of vivid orange feathers. The rest of the body is marked in the same way as the Adélie.

The mortality among the chicks being so very high, the probability is that the life of the adult is long, as otherwise the species could hardly survive. Dr. Herbert Klugh has calculated that the Emperor penguin lives for thirty-five years.

Evidence goes to show that the young birds spend their immaturity on the pack-ice, as all those sighted and collected on the pack at any distance to the northward have been immature, and no immature birds have been seen along the coasts at any time during the summer.

The food of the Emperor mainly consists of fish and crustaceans. There are invariably many small pebbles in the stomach. Like Adélies they must of course have open water within reach in order to get food, and in the neighbourhood of Cape Crozier this is always to be found, as the rapid tide there keeps the sea from freezing in over a considerable area, so that probably they never have to walk more than a mile or two to get food.

The cry of the Emperor is very loud and travels far across the ice. When sledging over the sea-ice in the spring, in the neighbourhood of Cape Adare, a curious sound was heard at times, reminding one strongly of the "overtone" notes of a ship's steam horn. The sounds puzzled us at the time, but I think now that most probably they were made by Emperor penguins.

The egg of the Emperor is white, pyriform in shape, and weighs just under 1 lb.

My own experience of these birds being limited I do not intend to enter deeply into the subject. The only surviving member of the band who visited Cape Crozier during the winter is Mr. Cherry-Garrard, and it is much to be hoped that some day he will write us an account of what he saw there. In the meantime for further details of the habits and morphology of the species, the reader is referred to Dr. Wilson's work, published in the second volume of the British Museum Reports, on the National Antarctic Expedition 1901–1904.

---

(1) *Pygoscelis adeliæ*.

(2) Sea-Leopard = *Stenorhinchus leptonyx*.

(3) Fig. 12.

(4) Fig. 37.

(5) This "watch-bill" was kindly kept for me by Mr. Priestly on his meteorological rounds, the nests being near the thermometer screen.

(6) Fig. 62.

(7) Fig. 70.

(8) This was invariable at Cape Adare.

(9) *Aptenodytes forsteri*.